2013

万达商业规划

持有类物业　上册 VOL.1

WANDA COMMERCIAL PLANNING 2013
PROPERTIES FOR HOLDING

万达商业规划研究院　主编

中国建筑工业出版社

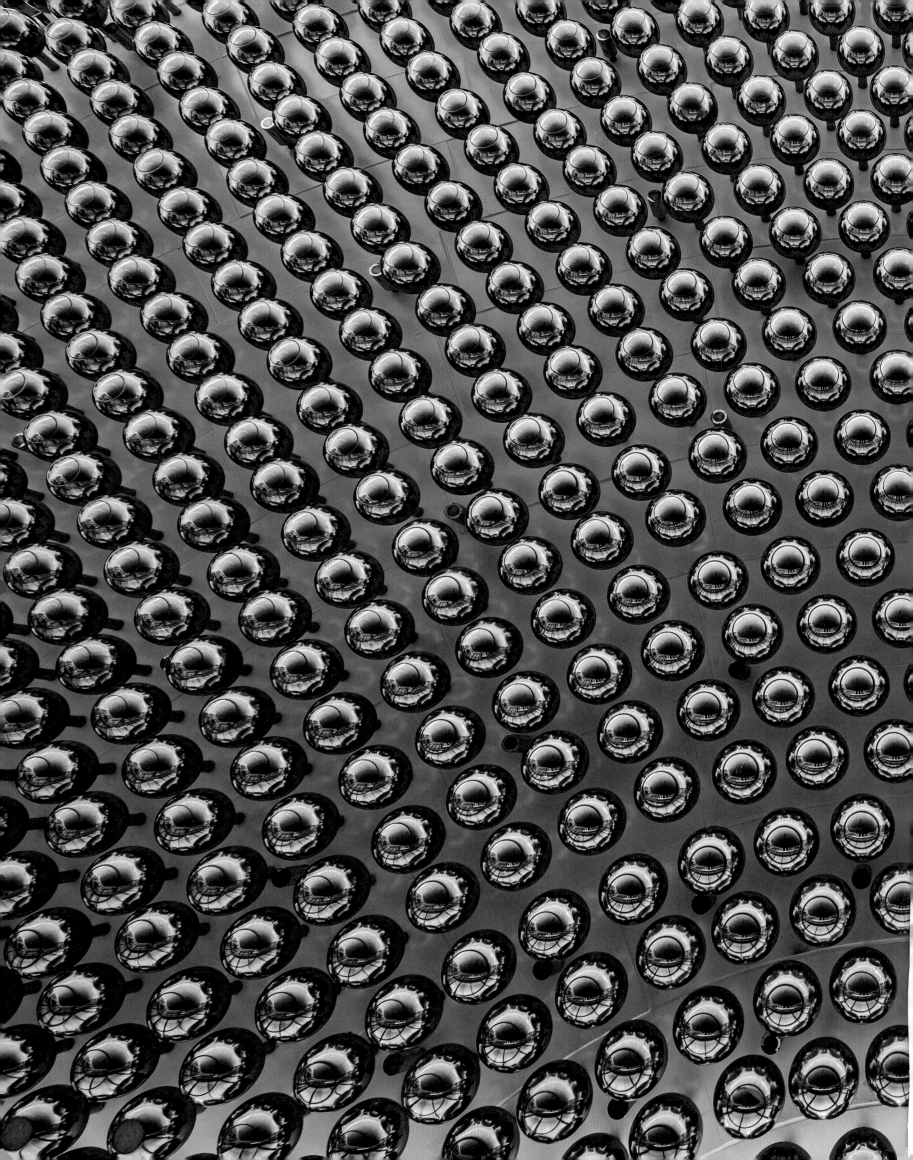

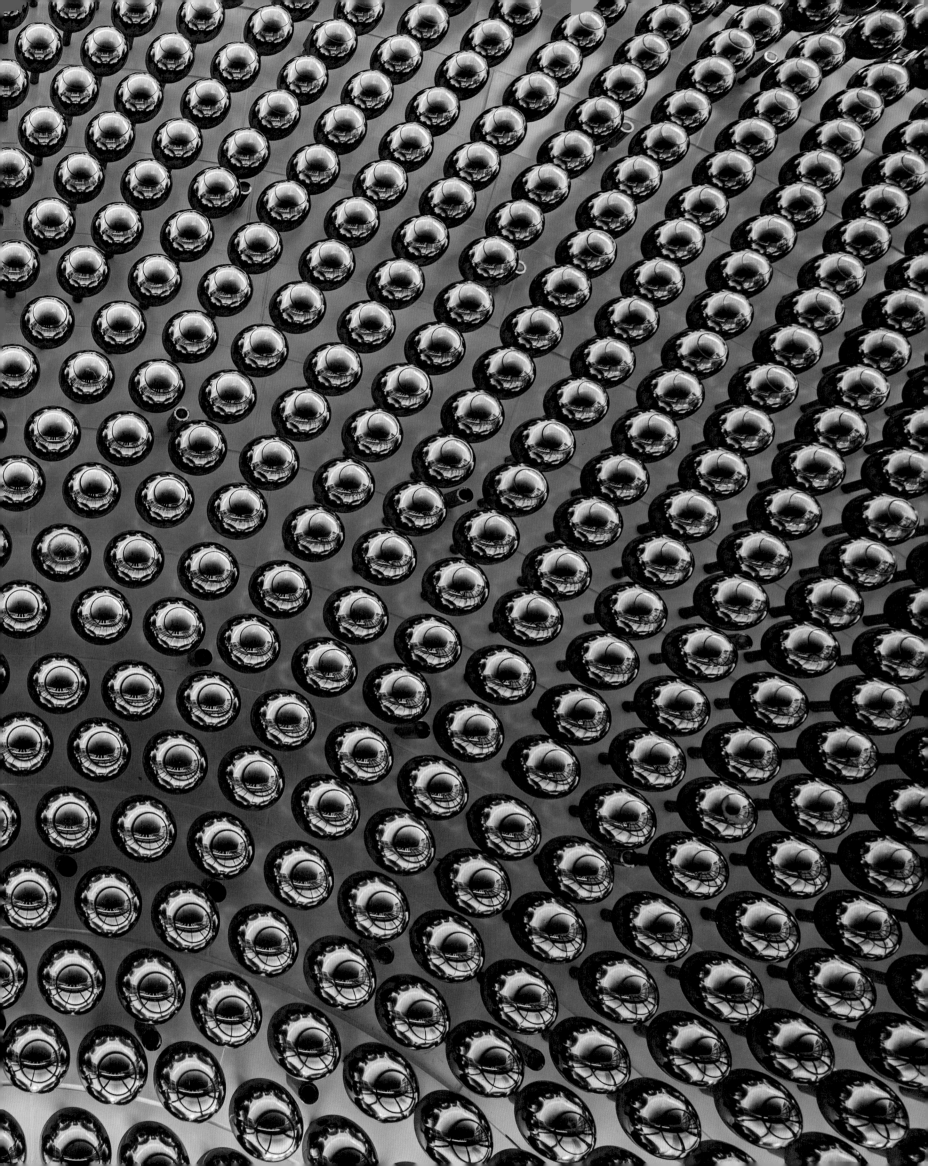

CONTENTS
目录

PART A FOREWORD
序言　010

012 / **QUALITY IMPROVEMENT MEETS EXPECTATION WITH MAJOR BREAKTHROUGH OF GREEN BUILDING AND HUIYUN SYSTEM (SMART-CLOUD SYSTEM)**
品质提升达到预期　慧云绿建重大突破

014 / **WANDA COMMERCIAL PLANNING 2013 - INNOVATION AND DEVELOPMENT OF WANDA PLAZAS**
万达商业规划 2013——万达广场的创新与发展

PART B WANDA PLAZAS
万达广场　020

022 / **WUHAN HAN STREET WANDA PLAZA**
汉街万达广场

054 / **DALIAN HIGH-TECH WANDA PLAZA**
大连高新万达广场

084 / **YIXING WANDA PLAZA**
宜兴万达广场

122 / **SHENYANG OLYMPIC WANDA PLAZA**
沈阳奥体万达广场

158 / **XIAMEN JIMEI WANDA PLAZA**
厦门集美万达广场

182 / **WUXI HUISHAN WANDA PLAZA**
无锡惠山万达广场

204 / **DONGGUAN CHANG'AN WANDA PLAZA**
东莞长安万达广场

234 / **CHANGCHUN KUANCHENG WANDA PLAZA**
长春宽城万达广场

260 / **HARBIN HAXI WANDA PLAZA**
哈尔滨哈西万达广场

PART C INDEPENDENT HOTELS
独立酒店 282

284 / **WANDA REALM WUHAN**
武汉万达嘉华酒店

298 / **WANDA VISTA TIANJIN**
天津万达文华酒店

314 / **WANDA REALM NANCHANG**
南昌万达嘉华酒店

326 / **WANDA REALM YINCHUAN**
银川万达嘉华酒店

PART D DESIGN & CONTROL
设计及管控 340

342 / **DEVELOPMENT HISTORY OF INTERIOR QUALITY IMPROVEMENT OF WANDA INDOOR PEDESTRIAN STREET**
万达广场室内步行街内装品质提升的沿革

346 / **DESIGN AND CONTROL OF WANDA PLAZA GUIDE SIGNAGE SYSTEM**
万达广场导向标识设计与管控

350 / **DEVELOPMENT HISTORY OF WANDA PLAZA LANDSCAPE DESIGN**
万达广场景观设计历史沿革

QUALITY IMPROVEMENT MEETS EXPECTATION WITH MAJOR BREAKTHROUGH OF GREEN BUILDING AND HUIYUN SYSTEM (SMART-CLOUD SYSTEM)

品质提升达到预期　慧云绿建重大突破

2013年是万达集团"三年品质提升年"的最后一年。评审小组反映，他们到各地公司最大的感受，是从公司领导到员工追求品质的精神，大家共同对各方面品质提升提建议、想办法。这说明万达"品质提升年"的目的已经达到。我们搞三年品质提升，不仅为了把广场做漂亮，更重要的是提升全员品质意识，树立品质第一的思想。

2013年，汉街万达广场、长沙开福万达广场、厦门集美万达广场、沈阳万达文华酒店、长沙万达文华酒店、长白山柏悦酒店、长白山凯悦酒店等一批优秀项目开业，大幅提高万达产品的美誉度。万达产品美誉度从哪里来？就是产品一个比一个好，长期积累而来。万达一年开业十几个酒店，在建施工酒店六十多个，术业有专攻，几年下来他们当然成了行业内的大专家。

万达商业规划研究院历经一年时间，在信息管理中心支持下，研发出万达广场慧云智能管控系统。这个系统把万达广场的消防、节能、运营、监控等几大功能集合在一个屏幕上，完全实现自动化、高科技管理。我到现场看了演示，感到非常高兴，给这个系统取名为"慧云"，寓意智慧的云。信息时代的管理，技术层面就是云计算，应用层面就是大数据，万达将来有几百个万达广场，上百家酒店，管理必须靠云计算、大数据。慧云系统由于刚刚研发出来，去年只在4个广场正式上线，但反响很好。我给它的评价是三个大幅提高：大幅提高安防水平、大幅提高节能水平、大幅提高管理水平。过去消防监控靠人盯，但人总有犯困的时候，总有偷懒的时候，慧云系统完全解决人为因素影响。过去万达广场消防、安全、运营监控分散在不同区域，不好管理，现在全部结合在一起，便于集中管理。我希望万达商业管理和万达项目系统的同志都要好好学学慧云系统。

2013年开业的全部万达广场、酒店均获得绿建设计认证，18个万达广场、7个酒店获得绿建运行认证，其中莆田万达广场获得绿建二星运行认证，是全国唯一获得绿建运行二星认证的商业项目。绿建运行认证比设计认证更难，万达一家企业获得的绿建运行认证占全国的80%，说明万达环保工作远远走在全国企业前面。

——摘自王健林
《万达集团2013年工作总结暨2014年工作安排》

2013 is the third and final year of Wanda's Quality Improvement Initiative. The review panel, when visiting project companies, was deeply impressed by the high spirit of all local employees for pursuing quality, with suggestions and solutions to improve overall quality. It shows that Wanda has achieved the purpose of the three-year quality improvement. We carried out the three-year quality improvement not only for beautiful plazas, but what's more important is to enhance the full consciousness of quality and establish the sense of treating the quality as top priority.

In 2013, the opening of a series of excellent projects greatly improved the reputation of Wanda products, including Wuhan Han Street Wanda Plaza, Changsha Kaifu Wanda Plaza, Xiamen Jimei Wanda Plaza, Wanda Vista Shenyang, Wanda Vista Changsha, Park Hyatt Changbaishan and Hyatt Regency Changbaishan. Wanda has won its reputation through long-term accumulation of ever better products. For the past several years, Wanda has opened more than 10 hotels each year, with over 60 hotels under construction. With the accumulated experiences and expertise, Wanda has become the leader of the industry.

Supported by Group IT Centre, Wanda Commercial Planning & Research Institute has spent one year to develop the Wanda Plaza Huiyun Intelligent Control System. This system integrates the functions of fire control, energy saving, operation and monitoring into one screen to fully implement automatic control and high-tech management. I was very delighted when I saw the demonstration on the site. The system is given a name Huiyun, meaning smart cloud. In the information era, management technically relies on cloud computation through applying big data. In the future, Wanda has to count on cloud computation and big data to manage its assets of several hundred Wanda plazas and over one hundred hotels. As a startup, the Huiyun System had just been applied to 4 Wanda plazas last year, but with very positive feedbacks. I described the system with three major advancements: 1) greatly improved the security-alert standard; 2) greatly improved the energy saving standard; 3) greatly improved the management standard. The fire control used to rely on manual monitoring, under which circumstances human flaws such as doziness and laziness are unavoidable, but can be completed addressed through Huyun System. The monitoring over fire control, safety/security and operations used to be separated, creating difficulties to management. The Huiyun System has integrated all

these tasks into centralized management. I encourage colleagues of Wanda Commercial Management and Wanda Project System to have a clear understanding of the Huiyun System.

All the Wanda plazas and hotels opened in 2013 were granted Green Building Design Labels. 18 Wanda plazas and 7 hotels were granted Green Building Operation Labels, among which Putian Wanda Plaza was granted two-star Green Building Operation Label, which is the only national commercial project granted two-star Green Building Operation Label. It's more difficult to get the Green Building Operation Label than the Green Building Design Label. Wanda's Green Building Operation Label counts for 80% of the nationwide labels, posing Wanda as the leading role of environment protection among all Chinese enterprises.

Quoted from Annual Work Report 2013 And Work Plan 2014 of Wanda Group
By Chairman Wang Jianlin

万达集团董事长
王健林
Wang Jianlin
Chairman of Wanda Group

WANDA COMMERCIAL PLANNING 2013 - INNOVATION AND DEVELOPMENT OF WANDA PLAZAS

万达商业规划 2013
——万达广场的创新与发展

文 / 大连万达商业地产股份有限公司高级总裁助理
兼万达商业规划研究院院长　赖建燕

2013年，万达商业地产开业万达广场18座，酒店16个。截止到2013年底，万达共开业85座万达广场，54个酒店，万达自持物业面积达到1704万平方米，位列世界第二。

◇◇◇◇◇◇◇◇◇◇◇◇◇◇◇◇◇◇◇◇◇◇◇◇◇◇

2013年也是万达商业地产总部迁京的第六个年份，回顾六年来，万达商业规划伴随商业地产的发展经历了五年不同的发展历程，如果每年需用一个关键词来简单归纳的话，我们不妨概括地总结一下。

2008年为万达商业规划的基础年
2008年万达商业规划建立了项目中心规划设计部，与商业规划院一起，形成了万达商业规划对销售与持有两大物业的规划管理模式，同年，完成了集团首个完整的商业规划标准《万达商业规划设计准则》。

2009年为万达商业规划的发展年
2009年当年开业8个万达广场和2个五星级酒店；之后的2010年当年开业万达广场15个，五星级酒店5个，使2009年商业规划的项目设计及管控数量较2008年骤增三到四倍（图1）。

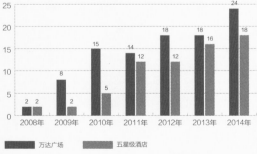

(图1) 万达商业规划发展

2010年为万达商业规划的标准化年
2010年，万达商业规划研究院共组织制定和颁布实操性集团技术标准44项，参与了2010版《万达集团制度》制定，并主持完成了《万达2010版建造标准》、《万达购物中心节能工作指南》、《消防性能化标准及工作指南》、《万达幕墙设计管控标准》和《地下停车场导向标识设计标准》等集团标准17项；2010年也是万达**绿建节能年**。万达商业规划研究院牵头完成的《万达集团"绿色、低碳"战略研究报告》，首次公开明确地提出了万达集团商业建筑"绿色、低碳"战略目标，即：2011年及以后开业的

In 2013 Wanda Commercial Real Estate opened 18 plazas and 16 hotels. By the end of 2013 Wanda owned 85 Wanda plazas and 54 hotels, which has total holding property area of 17,040,000 m^2, ranking the second in the world.

◇◇◇◇◇◇◇◇◇◇◇◇◇◇◇◇◇◇◇◇◇◇◇◇◇◇

The year 2013 is also the sixth year after Wanda Commercial Real Estate had relocated its headquarters to Beijing. Looking back over the past 6 years, Wanda Commercial Planning experienced 5 years of different developing courses of the commercial real estate. Some keywords may simply summarize each past year as below.

2008 IS THE YEAR OF ESTABLISHMENT FOR WANDA'S COMMERCIAL PLANNING
In 2008, Wanda Commercial Planning set up the Planning and Design Department of the Project Center, which, together with the Commercial Planning Institute, formed the planning management mode of Wanda Commercial Planning for selling and holding property. In the same year, Wanda Group established its first complete commercial planning standard: *Wanda Commercial Planning and Design Standards*.

2009 IS THE YEAR OF DEVELOPMENT FOR WANDA'S COMMERCIAL PLANNING
In 2009, eight Wanda plazas and two 5-star hotels were opened. In 2010, 15 Wanda plazas and five 5-star hotels were opened, the number of commercial planning design and control projects in 2009 has increased to 3 to 4 times compared to projects in the year 2008 (Fig. 1).

2010 IS THE YEAR OF STANDARDIZATION FOR WANDA'S COMMERCIAL PLANNING
In 2010, Wanda Commercial Planning Institute organized the formulation and promulgation of 44 Wanda practical technical standards, participated in the development of *Wanda Group Regulation* 2010 Version, and were responsible for the completion of 17 group standards, including *Wanda Construction Standards 2010 Version, Energy Saving Guidelines of Wanda Shopping Centre, Performance-based Fire Control Standards and Instructions, Wanda Curtain Wall Design and Control Standards* and *Design Standards of Underground Parking Signage*. 2010 is also **Wanda's year of green building and energy saving**. Wanda Commercial Planning Institute presided the *Completion of "Green and Low Carbon" Strategy Research Report of Wanda Group*, and for the first time explicitly put forward the "green and low carbon" strategic objectives of Wanda commercial

项目均取得绿色建筑一星设计标识；2011年至2015年间新开业项目逐年降低运行能耗2%~3%；2013年2个项目取得绿色建筑一星运行标识认证；2015年实现运营管理水平均达到绿色建筑一星运营标准。之后，万达全面兑现承诺，并完成了中国商业建筑"绿色建筑设计标识"、"绿色建筑运营标识"零的突破（图2）。

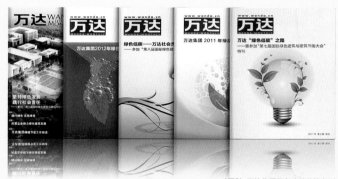

（图2）万达集团绿色建筑节能专刊

2011年为万达商业规划的品质年

2011年，大连万达商业地产股份有限公司首次对当年开业的万达广场进行了"品质"排名评比。同年万达商业规划研究院也组织完成了万达广场品质提升共5类108项工作。

2012年为万达商业规划的成本年

2012年，万达商业规划研究院、项目中心设计部与集团成本部等部门历时一整年，共同完成了《万达广场定额设计技术标准》及《万达设计成本管控标准及操作指引》，使得规划设计更加融入商业地产项目操作整体链条。《万达广场定额设计技术标准》及《万达设计成本管控标准及操作指引》的研讨和标准的建立，使万达商业规划体系与各系统、部门、公司建立了共同的操作标准、评判标准；使项目的成本从规划设计这一根本源头达到整体控制。

万达商业规划十二年的历程记录在《万达商业规划2009~2012》的图目年鉴中。万达商业规划是在万达集团王健林董事长的亲自指导下建立、成型和发展的，是在丁本锡总裁的支持下多年不断持续的总结、积累和完善的。商业地产核心产品规划设计的整体策划与创新直接来自王健林董事长对商业地产模式的创新与把控；商业规划管理体系的宏观构思是

buildings: the projects opened in 2010 and later should be all granted One-Star Green Building Design Label; new projects opened between 2011 and 2015 should reduce the energy consumption by at least 2% to 3% every year; Two One-Star Green Building Operation Labels should be granted in 2013; and in 2015 the management level of all the projects should achieve One Star Green Building Operation Standards. Later, Wanda fulfilled the commitments and achieved the zero breakthrough of commercial buildings' Green Building Design Label and Green Building Operation Label in China (Fig. 2).

2011 IS THE YEAR OF QUALITY FOR WANDA'S COMMERCIAL PLANNING

In 2011, for the first time Dalian Wanda Commercial Real Estate Co., Ltd. Started to rank the "quality" of Wanda plazas opened in that year. In the same year, Wanda Commercial Planning Institute organized 5 categories and 108 items to complete the quality improvement of Wanda plazas.

2012 IS THE YEAR OF COST CONTROL FOR WANDA'S COMMERCIAL PLANNING

In 2012, Wanda Commercial Planning & Research Institute, Project Centre Design Department and Cost Department of Wanda Group spent one year completing *Technical Standards of Quota Design for Wanda Plazas* and *Wanda Design Cost Control Standards and Operating Guidelines*, integrating planning and design more deeply into the commercial real estate projects through the whole operational chain. The research and establishment of the above standards unified the operation standards and evaluation standards for Wanda Commercial Planning System and other different systems, departments and companies; thus the project cost is overall controlled from the starting source of planning and design.

The 12-year history of Wanda Commercial Planning has been recorded in the yearbook chart of *Wanda Commercial Planning 2009-2012*. Wanda Commercial Planning is established, formed and developed under the personal guidance from Chairman Wang Jianlin, and is continuously summarized, accumulated and improved for many years under the support of President Ding Benxi. The overall planning and innovation for the planning and design of the core commercial real estate product is directly conceived from the innovation and control by Chairman Wang Jianlin; and the macro conception of the commercial planning and management system is a part of the complete system of Wanda commercial real estate project control. Five years after Wanda had relocated

万达商业地产项目管控完整体系的一部分。万达总部进京的五年，为2013年万达商业规划的全程信息化管理以及全面的学术总结奠定了坚实的基础。

◇◇◇◇◇◇◇◇◇◇◇◇◇◇◇◇◇◇◇◇◇◇◇◇◇◇◇

2013年可以用三个关键词来概括，即**创新、信息化和学术总结**。

2013年，是万达商业规划深度创新的一年

2013年建成开业的"汉街万达广场"、"长沙开福万达广场"可以说代表了迄今开业的万达广场商业规划的最高水平。无论是购物中心的整体形象，还是平面商业规划的功能及动线布局，都突破了以往万达常态化的既有模式，有所创新，成为已开业85个万达广场中的标志性项目（图3、图4、图5）。

（图3）汉街万达广场

2013年万达商业规划的创新主要表现在以下几个方面：首先是总图规划呈现布局的灵活多样。经过多年的探索，尤其是市场对已开业几十个万达广场的反馈，2013年万达商业规划的总图布局在尊重环境与控规的基础上，更加灵活且更加贴近市场。万达商业规划研究院引入内部竞争机制，所有项目的概念规划设计需通过多方案竞赛选取，在管理上确立了创新的机制。

第二是平面设计根据市场反馈而调整。2013年，万达商业规划根据万达商业管理公司多年的市场分析反馈，对购物中心体验业态与零售的比例进行了系统的调整，对百货与室内步行街的平面动线进行了水平与竖向两个方向的优化，进一步提高了购物中心的运营平效。

第三是形象设计体现地方文化的主题。不论是立面设计、内装设计还是景观设计及夜景照明设计，2013年万达商业规划在效果类设计上，更加强调与地方文化相结合，更加强调设计的文化主题。当年开业的厦门集美万达广场体现了当地的"嘉庚建筑风格"，并在景观和内装上展示了"惠安"的文化主题；长春宽城万达广场展示了"电影之城"的文化主题；2013年，大连高新万达广场、南京江宁万达广场、沈阳奥体万达广场、东莞长安万达广场、

the headquarter in Beijing, we have built a solid foundation for the whole information management as well as the comprehensive academic summary of Wanda Commercial Planning in 2013.

◇◇◇◇◇◇◇◇◇◇◇◇◇◇◇◇◇◇◇◇◇◇◇◇◇◇◇

The year 2013 can be described with three keywords: **innovation**, **informationization** and **academic summary**.

THE YEAR 2013 WITNESSED THE DEEP INNOVATION OF WANDA COMMERCIAL PLANNING

Wuhan Han Street Wanda Plaza and Changsha Kaifu Wanda Plaza, which were built and opened in 2013 can be said to represent the highest commercial planning level of all previous Wanda plazas. Their overall shopping center images as well as the functions and dynamic layout of the commercial planning plan, and

（图4）长沙开福万达广场

surpassed the entire previous Wanda standard model and became the benchmark project of 85 opened Wanda plazas (Fig. 3, Fig. 4 and Fig. 5).

The innovation of Wanda Commercial Planning in 2013 is mainly represented in the following aspects: Firstly, the general planning presented flexible and various layouts. After years of exploration, especially the market feedbacks from dozens of opened Wanda plazas, in terms of the environment and regulations, the general layout of Wanda Commercial Planning in 2013 was more flexible and closer to the market. Wanda Commercial Planning Institute introduced internal competition mechanism. The conceptual planning and design for every project was selected from multiple scheme competitions, which established the innovative mechanism of management.

Secondly, the plan design was adjusted in accordance with the market feedbacks. In 2013, Wanda Commercial Planning carried out systematic adjustment to the proportions of the experience programs and the retail, in accordance with years of market analysis and feedbacks from Wanda Commercial Management Corporation, and optimized the horizontal and vertical circulations of, the department store and the indoor pedestrian street, which further improved the operational efficiency of the shopping center.

Last but not the least, the facade design reflected

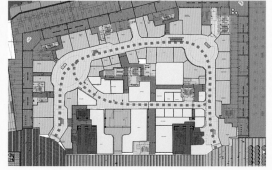

（图5）长沙开福万达广场动线图

丹东万达广场等众多的万达广场都表现出了极具个性化的创新。汉街万达广场的夜景照明设计，获得了2014年度第31届IALD国际照明设计大奖，成为中国首个问鼎IALD奖项的商业地产项目，也是中国首次有项目获得IALD建筑照明类最高级别奖项——卓越奖(图6)。(编者注：IALD是IALD国际照明设计奖(International Association of Lighting Designers Awards)的缩写 ，一年一届，是全球照明行业的最高级别奖项，享有全球照明界"奥斯卡"美誉，代表了当年全球最高设计水平)

the local culture. The visual effect design of Wanda Commercial Planning in 2013 emphasized more on the integration of the local culture and the theme of culture, which includes exterior facade design, interior design, landscape design and nightscape lighting design. Xiamen Jimei Wanda Plaza, Which was opened in 2013, reflected the local "Jiageng architectural style" and in its landscape and interior design both presented the "Huian" cultural theme. Changchun Kuancheng Wanda Plaza also presented the cultural theme of "Film City". In 2013, a series of Wanda plazas showed highly individualized

（图6）汉街万达广场夜景

2013年，是万达商业规划信息化全面实施的一年
万达商业规划**信息化**的全面实施，有两个主要标志。第一是实现万达商业规划设计系统的全程信息化管控。2013年，商业规划设计系统实现了OA、计划模块与图文档系统三个信息化平台的对接与连动，成为集团首个实现全程信息化管控的业务系统；并与信息中心合作，成功为图文档系统申请了国家"计算机软件著作权"（图7、图8）。2013年，商业规划设计系统对集团2000~2017年已竣工的85个万达广场、55个酒店和129个在建、在设大商业项目、酒店项目以及148个住宅项目的设计成果进行了分目归档，共形成20779个条目，51075份档案。实现了规划图文档

innovation, including Dalian High-Tech Wanda Plaza, Nanjing Jiangning Wanda Plaza, Shenyang Olympic Wanda Plaza, Dongguan Chang'an Wanda Plaza and Dandong Wanda Plaza. The nightscape lighting design of Han Street Wanda Plaza won the 31st IALD International Lighting Design Award in 2014. It is the first commercial real estate project of China winning IALD award and also the highest award of construction lighting design - IALD Award of Excellence(Fig. 6).(Editor's note: IALD, stands for International Association of Lighting Designers Awards, which held once a year, is the highest award in the global lighting industry and has the reputation of "Oscar in Global Lighting". It represents the world's highest design level of the year)

THE YEAR 2013 WITNESSED THE FULL MPLEMENTATION OF INFORMATIONIZATION OF WANDA COMMERCIAL PLANNING
There were two main signs of the full implementation of **informationization.**

The first sign was the achievement of the whole-process information control of Wanda Commercial Planning and Design system. In 2013, the commercial planning and design system achieved the docking and linkage of the three informationized platforms - OA, project

（图7）"计算机软件著作权"登记证书

（图8）万达规划设计图文档管理系统界面

系统对设计档案进行全面收集、整理的开发目标。2013年, 商业规划设计系统全面完成"制度"、"标准"、"强条"与管控信息化平台的统一整合, 增强了系统的实用性。

第二是万达广场慧云智能化管理系统试点项目的运行成功。2013年, 由万达商业规划研究院主导, 与万达商业管理有限公司等公司共同成功研发了万达广场慧云智能化管理系统, 完成四个试点工程的验收, 并成功申请了国家"计算机软件著作权"。该系统的研制成功并上线实施, 标志着万达在安全、节能管理方面已成为引领行业的先驱(图9)。为保证慧云智能化管理系统的实施与使用效果, 万达商业规划研究院还编制了《2012年开业项目弱电智能化状况白皮书》, 对集团2012年及2013年开业项目的弱电智能化系统的实施状况进行了全面总结和诊断, 确保了慧云系统的实施。

(图9) 万达广场慧云智能化管理系统界面

2013年, 是万达商业规划进行学术总结提升的一年
万达集团2000年进军商业地产, 2013年首度正式著述对外出版发行, 万达商业规划研究院作为万达商业地产股份有限公司的主要研发部门, 参与了万达商业地产股份有限公司编著两本图书:

一本是大连万达商业地产股份有限公司著, 王健林董事长主编, 由清华大学出版社2013年12月出版的《商业地产投资建设》(图10); 另一本是由万达商业规划研究院、万达商业管理有限公司联合编著, 王健林董事长作序, 由中国建筑工业出版社2013年6月出版的《绿色建筑——商业地产中绿色节能的实践及探索(一)》(图11)。

此外, 2013年万达商业规划研究院还对截止到2012

schedule module and drawings and documents archive system, and became Wanda's first operation system with full implementation of the whole-process information control. The cooperation with the Information Center assisted the drawings and documents archive system to obtain the national "copyright of intelligent computer software" (Fig. 7 and Fig. 8). In 2013, the commercial planning and design system has archived Wanda's completed 85 Wanda plazas and 55 hotels, and also 129 commercial and hotel projects and 148 residential projects under construction, from 2000 to 2017, and formed 20,779 archive entries and 51,075 files. It achieved the development objective of the drawings and documents archive system to comprehensively collect and collate the design archive files. In 2013, the commercial planning and design system completed the integration of "Regulations", "Standards", "Compulsory Requirement" and the informationized control platform, enhancing the practicability of the system.

The second sign was the operation success of Huiyun Intelligent Management System on the pilot projects. In 2013 Wanda Huiyun Intelligent Management System, which was jointly developed by Wanda Commercial Management Co., Ltd. under the lead of Wanda Commercial Planning Institute, completed the acceptance of four pilot projects and successfully obtained the national "copyright of intelligent computer software". The successful development and launch of the system marked that Wanda had become a pioneer leader of the industry in terms of safety and energy saving management (Fig. 9). Wanda Commercial Planning Institute developed *Intelligent Low-voltage Electrical Situation of Projects Opened in 2012*, which fully reviewed and diagnosed the implementation of the intelligent low-voltage electrical system in the projects opened in 2012 and 2013, to ensure the implementation of Huiyun Intelligent Management System.

THE YEAR 2013 ALSO WITNESSED THE ACADEMIC SUMMARY IMPROVEMENT OF WANDA COMMERCIAL PLANNING

Wanda Group entered commercial real estate in 2000, and published the first official academic summary in 2013. As one of the main developing departments of Wanda Commercial Real Estate Co., Ltd., Wanda

(图10)《商业地产投资建设》　　(图11)《绿色建筑——商业地产中绿色节能的实践及探索(一)》

年底万达商业地产已开业的67个万达广场、38个五星级酒店等自持物业进行了梳理总结,主编了分别于2013年6月由中国建筑工业出版社出版《万达商业规划2009~2011》的实录年鉴,于2013年12月出版《万达商业规划2012》的实录年鉴(图12)。

至此,万达商业规划的学术总结进入制度化的轨道。今后的每年上半年,我们都将总结出版上一年的《万达商业规划实录年鉴》。本次出版的《万达商业规划2013》实录年鉴,将首次分万达商业规划持有类物业和销售类物业两大部分的内容,使万达商业规划图文档案更加及时完整。

(图12)《万达商业规划》系列丛书

2013年,万达商业规划系统进行了调整,2月,万达商业规划研究院文化旅游分院独立为万达文化旅游规划研究院,隶属北京万达文化产业集团有限公司。

万达商业规划系统的工作,得到了万达集团王健林董事长的亲自指导和关怀,至今万达商业地产的所有发展项目的总图、平面功能及建筑效果均需经董事长审图会确认方可执行。万达商业规划系统的工作是万达整体产业链的一部分,商业规划管理体系的健全也离不开万达集团丁本锡总裁、万达商业地产股份有限公司齐界总裁的指导!在此,对所有支持、帮助过万达规划系统的所有领导、同事、朋友表示诚挚的谢意,也对万达集团各系统、部门、项目公司的支持表示衷心感谢!

Commercial Planning Institute has participated in editing two books as below.

The first book is *Commercial Real Estate Investment and Construction* (Fig.10), which was edited by Wanda Commercial Real Estate Co., Ltd., and Chairman Wang Jianlin is the editor-in-chief. The book was published by Tsinghua University Press in December 2013. The second book is Green Buildings - *Green and Energy Saving Practice and Exploration in Commercial Real Estate (I)* (Fig. 11), jointly edited by Wanda Commercial Planning & Research Institute and Wanda Commercial Management Co., Ltd., the preface was written by Chairman Wang Jianlin, and the book was published by China Architecture & Building Press in June 2013.

In addition, Wanda Commercial Planning Institute has summarized the holding properties of Wanda Commercial Real Estate which opened up to the end of 2012, including 67 Wanda plazas and 38 5-star hotels, and edited the yearbooks *Wanda Commercial Planning 2009 to 2011* and *Wanda Commercial Planning 2012* (Fig. 12) published by China Architecture & Building Press respectively in June 2013 and December 2013.

At this point, the academic summary of Wanda Commercial Planning entered the institutionalized track. In the first half of the year from now on, we will summarize and publish *Wanda Commercial Planning Yearbook* of the previous year. *Wanda Commercial Planning 2013* published this time is for the first time being separated into Wanda holding properties and Wanda selling properties, making the graphic files and documents of Wanda Commercial Planning more timely and complete.

Wanda Commercial Planning System was reassembled in 2013. In February, Cultural Tourism Branch of Wanda Commercial Planning Institute became an independent institute under Beijing Wanda Cultural Industries Group Co., Ltd.

The work of Wanda Commercial Planning System received the personal guidance and care from Chairman Wang Jianlin. The site plan, plane functions and architectural visual effects of the projects developed by Wanda Commercial Real Estate are all required to be reviewed and confirmed by the Chairman before the implementation. The work of Wanda Commercial Planning System consists of a key part of Wanda industry chain. The commercial planning and management system could not be completed without the guidance of President Ding Benxi of Wanda Group and President Qi Jie of Wanda Commercial Real Estate Co., Ltd. We would like to express the sincere gratitude to all the leaders, colleagues and friends who have supported and helped Wanda Planning System and the heartfelt appreciation to the support of each system, department and Project Company of Wanda!

重庆万州万达广场

重庆万州万达广场

审图号：GS（2014）1915号

哈尔滨哈西万达广场

长春宽城万达广场

抚顺万达广场

沈阳奥体万达广场

丹东万达广场

北京

大连高新万达广场

徐州云龙万达广场

西安大明宫万达广场

蚌埠万达广场

南京江宁万达广场

无锡惠山万达广场

宜兴万达广场

汉街万达广场

宁波余姚万达广场

长沙开福万达广场

赤尾屿

钓鱼岛

台湾岛

厦门集美万达广场

东莞长安万达广场

东沙群岛

海南岛

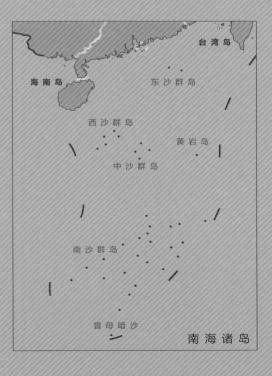

台湾岛

海南岛

东沙群岛

西沙群岛

黄岩岛

中沙群岛

南沙群岛

曾母暗沙

南海诸岛

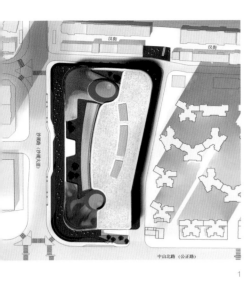

WUHAN HAN STREET WANDA PLAZA
汉街万达广场

开业时间	2013 / 09 / 28
建设地点	湖北 / 武汉
占地面积	3.39 公顷
建筑面积	13.45 万平方米

OPENED ON	SEPTEMBER 28 / 2013
LOCATION	WUHAN / HUBEI PROVINCE
LAND AREA	3.39 HECTARES
FLOOR AREA	134,500 m²

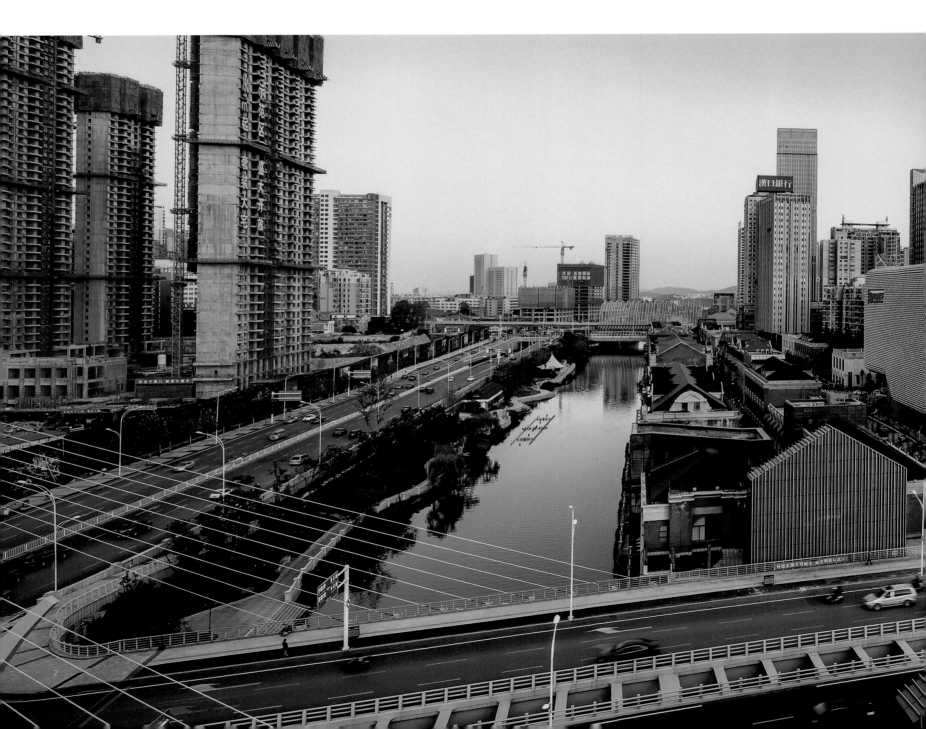

OVERVIEW OF PLAZA
广场概述

项目占地3.39公顷，总建筑面积达到13.45万平方米。项目在主要道路沙湖路方向设置两个广场及主入口，在东北角设置第三个入口与汉街相连，并利用与汉街近5米的高差将地下一层直接通向汉街。在靠近两个主入口处分别设置椭圆中庭及圆中庭，并在万达历史上第一次采用双街形式把两中庭联系贯通，形成了丰富的空间效果。广场由高端精品百货、电玩、大歌星KTV、万达影城旗舰店、高档餐饮、精品超市等业态组成。万达百货达5.5万平方米，首次突破主力店形式，以步行街精品商铺形式分布于一至三层；电玩和大歌星KTV设置在四层，与小餐饮业态充分融合，让吃喝玩乐形成一体化。特别指出的是，在项目的五层设置万达历史上第一个万达影城旗舰店，总面积达1.3万平方米共15个厅，精品超市设置于地下一层，面积仅为5800平方米，是万达历史上第一次设置精品超市。

This project covers total land area of 3.39 hectares with total floor area of 134,500 m². The project has two squares and two main entrances along with Shahu Road, and the third entrance locates at the northeast corner, which, taking the advantage of approximate 5-metre height difference on site, and connects B1 to Han Street. An elliptic atrium and a circular atrium are respectively arranged near the two main entrances, and for the first time in Wanda history "twin streets" are adopted to connect the two atriums, formed rich spatial effect. The shopping centre consists of high-end boutique department store, video games, Big Star KTV, Wanda Cinema flagship store, restaurants, supermarket and other commercial programs. Breaking through the form of anchor store for the first time, Wanda Department Store covers from ground floor up to the 3rd floor in the form of boutique shops in pedestrian street with total area of 55,000 m²; while video games and Big Star KTV are set on the 4th floor to fully combine with small food and beverage, to integrate the shopping, eating, drinking, leisure and entertainment all together. What is particularly worth mentioning is the first Wanda Cinema flagship store of Wanda Group on the 5th floor, including total area of 13,000 m² and 15 cinemas. Meanwhile, the boutique supermarket on B1 with area of only 5,800 m² is also the first boutique supermarket of Wanda history.

1 广场总平面图
2 广场全景图

FACADE OF PLAZA
广场外装

外立面亮点——风格独特，与当地文脉紧密结合，最具建筑特色。

外立面概念取自于武汉水文化，建筑外立面通过抛光不锈钢材质模仿水反射环境特点，并通过9种不同切割程度球体的有机排列形成水流流动效果。高贵性则采用具有手工感的雪花石材质进行充分的展示，这不仅加强了建筑本身的城市符号概念，而且也强化万达集团的形象。外立面金属球达43222个，国内首创仿雪花石和抛光不锈钢组合制作而成，已申请获得国家专利；同时在万达广场首次采用单元式幕墙系统设计，工厂模块化生产和现场安装来实现。特别在两个主要入口部位，采用悬挑钢桁架结构形式门头，悬挑跨度达22米，气势宏大，是万达广场最大入口悬挑项目。

Highlight of the facade - a unique style combined with the local cultural context, formed a vivid architectural features.

The idea of the exterior facade is conceived from the lake culture of Wuhan. The polished stainless steel of the building facade imitates water's characteristic of reflecting the environment, and the organic array of spheres with 9 different degrees of cutting generates the effect of water flow. The nobility of the building is fully presented by the alabaster material with handmade sense, which not only emphasizes the building's concept of city symbol, but also enhances the brand image of Wanda Group. The facade, containing up to 43,222 metal spheres, is made of the combination of Polished stainless sted and-imitated alabaster, which is first time used in China and applied for national patent; meanwhile, it is for the first time in Wanda history unitized curtain wall design is adopted and achieved by modular production and installation. Cantilevered steel truss structure is especially used for the door lintel of the two main entrances with the cantilever span of 22m forming the grand momentum, making this project have the largest cantilevered entrance among Wanda's projects.

3 广场立面图

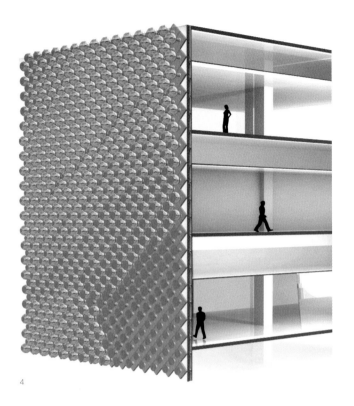

4

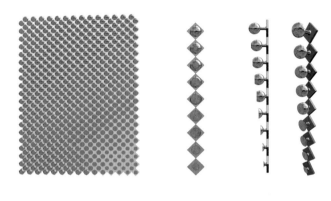

5

6

4　外立面金属球组合分析图
5　广场外立面金属球分布
6　广场立面效果

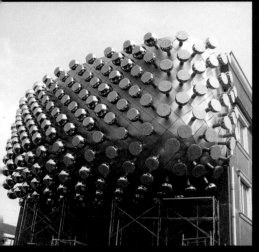

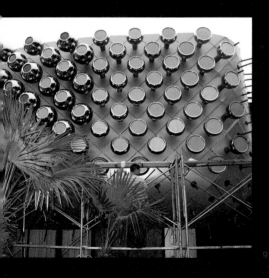

7 广场外立面模块特写
8 广场外立面模块仰视效果
9 广场外立面模块正面效果
10 广场外立面模块效果

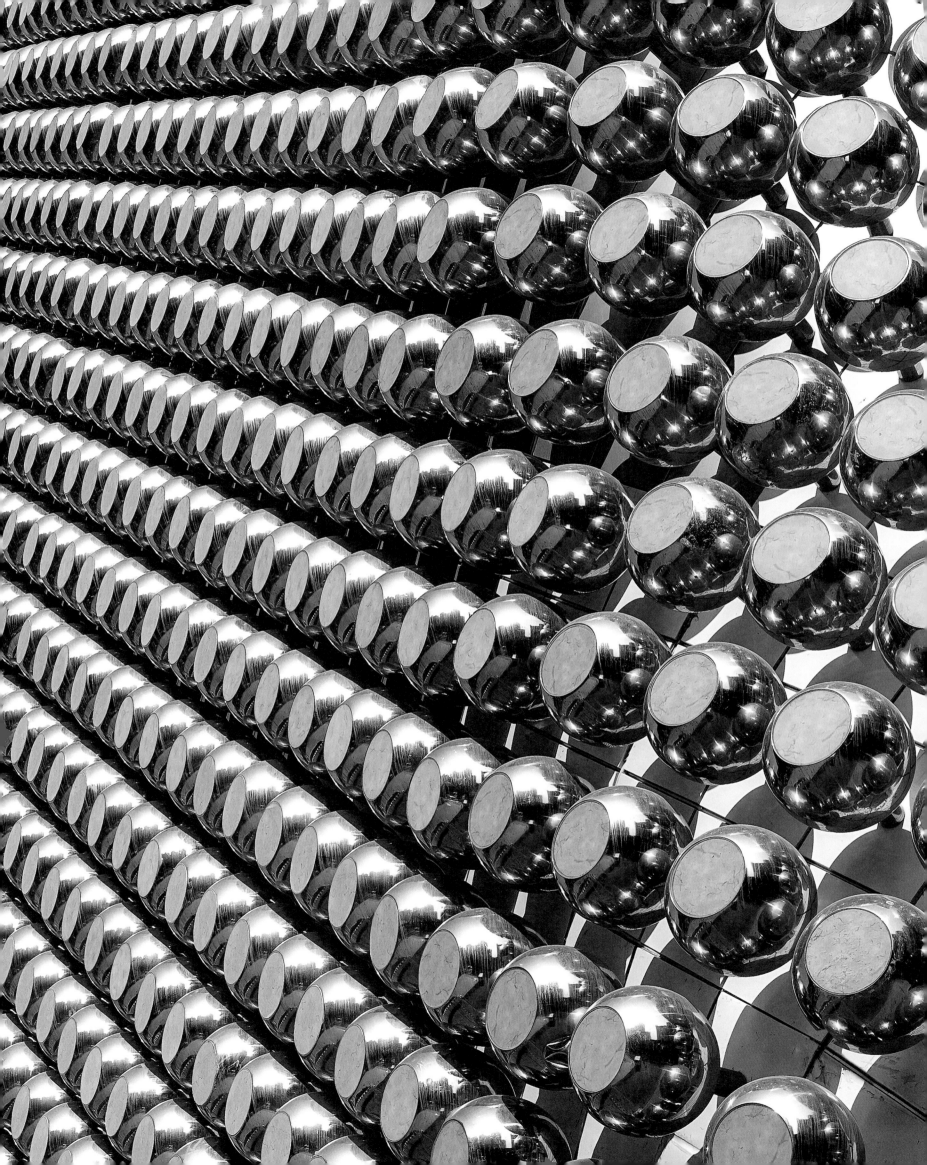

11

11 表皮模块顶视图
12 表皮模块透视图
13 广场西南主入口

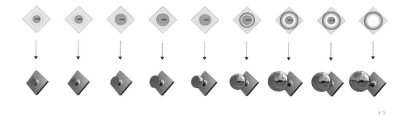

12

13

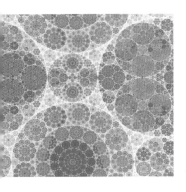

14a

14b

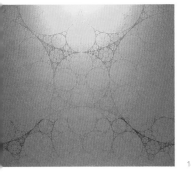

14c

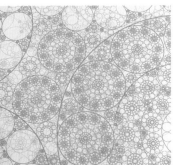

14d

14e

INTERIOR OF PLAZA
广场内装

汉街万达广场内装最核心的椭圆中庭与圆中庭, 采用双曲陀螺线结构造型, 把玻璃采光顶和观光电梯外围护结构形成一个整体, 子弹头观光电梯包裹其中, 结构优雅且具张力, 形成了室内整个空间的视觉焦点, 塑造出顶级奢侈品店的独特、奢华、时尚而具备特有的艺术气息。同时在结构造型表面以全彩数码打印玻璃披覆, 中间露出观光电梯, 使得梯内观众与中庭之间形成良好的空间互动关系。为防止雨水的渗入, 并提供足够的热工性能, 在漏斗形竖向网状立柱的顶部采用釉面玻璃幕墙屋盖将其封闭, 将雨水有机地引导至采光顶外围的屋顶天沟。

The oval hall and circular hall, which are core of Han Street Wanda Plaza, apply a structural form of hyperbolic spinning top to make the glass skylight and enclosure of panoramic lifts into one integral part, enclosing the bullet-shaped panoramic lifts, making the structure elegant and full of tension, and making it a visual focus of the overall interior space. By this means of design, it produces extraordinarily special artistic taste which is distinctive, luxurious and fashionable for the top luxury boutiques. On the surface of the structural form is covered with full-color digital glass film with the panoramic lifts exposing in the middle to form good spatial interaction between passengers in the panoramic lifts and the atrium. To prevent rain infiltration and to provide sufficient thermal performance, the design adopts ceramic glaze curtain wall cover to block the top of the vertical funnel-shaped mesh columns so that rain will be led to the roof gutter on the periphery of the skylight.

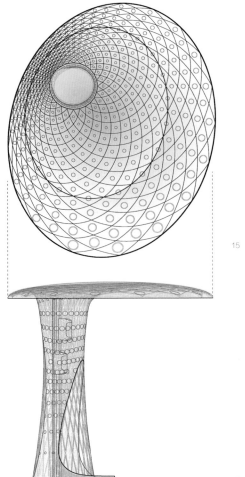

15

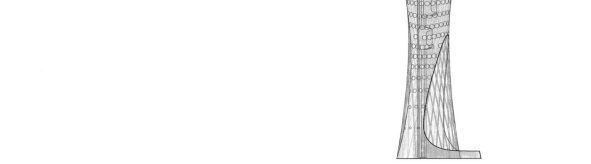

16

14 椭圆中庭玻璃图案
15 采光顶玻璃分割平面图
16 观光电梯造型玻璃分割立面图
17 椭圆中庭

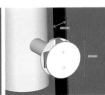

18 玻璃和抓点结构连接
19 椭圆中庭

20 图中庭电梯结构

步行街护栏在国内首次采用超大尺度无立柱悬臂式玻璃护栏体系，栏板玻璃总高2.2米，使得侧裙板更加整体，空间流动性更强，视线通透。中庭侧裙板在万达历史上首次采用三维多曲面蜂窝铝板，与中庭观光电梯钢结构相得益彰。地面石材拼花采用模数化水刀切割技术，图案精美自然而有机。在内装总体色调上两个中庭以及地面铺装的主色调定义为金色和银色——金色代表灿烂的历史文明，银色代表新兴的时尚文化；两个颜色巧妙地在长街变换融合，从而使整体空间更具活力和感染力。

The guardrail of pedestrian street has used large scale column-free cantilevered glass system for the first time in China. The glass of 2.2m high provides the integrity of the side panels, enhancing the space flow and the transparent sight. The side plates at the atriums adopt 3D multi-surface cellular aluminum panels, which has been used for the first time in Wanda, and they complement with the steel structure of the sightseeing elevators at the atriums. The stone floor pattern adopts modular water jet cutting technology, producing the exquisite, natural and organic patterns. In terms of the interior overall color, the main tone of the two atriums and the floor is defined as gold and silver-gold stands for the brilliant history of the civilization while silver stands for the emerging fashion culture. The ingenious transformation and infusion of the two colors strengthens the vitality and appeal of the whole space.

21

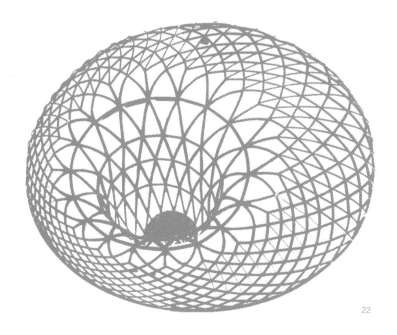

22

21 圆中庭仰视图
22 圆中庭模型
23 圆中庭

25

26

24 室内步行街
25 室内步行街地面图案
26 室内步行街空间

28

29

30

LANDSCAPE OF PLAZA
广场景观

汉街万达广场的建筑设计和景观设计都致力在设计上体现奢侈品的特色。景观设计从一开始就要求显示出定制的感觉，对场地设计的每个角落都与建筑和内装设计的主题相呼应。例如场地北侧的车行挡墙，其墙体的圆环铺装设计与地面铺装和建筑立面球体相呼应。尤其是地面铺装中的金属环配合室内的金厅和银厅位置而采用了不同的铜环和不锈钢环，实现室外细节定制一体化。甚至包括室外的防尘垫也要求生产厂家定制，与主题相呼应。

The architectural and landscape design of Han Street Wanda Plaza are committed to reflecting the characteristic of luxury. The landscape design was required at the very beginning to present the customized feeling, and every corner of the site should echo the theme of the architecture as well as the interior design. For example, the retaining wall of the roadway on the north of the site adopts ring cladding pattern which echoes the floor pavement and the facade spheres. Different copper rings and stainless steel rings are especially paved on the floor to emphasis the gold atrium and silver atrium and realize the integration of customized details from interior to exterior. Even the exterior dust mats are customized by the manufacturers to echo the theme pattern.

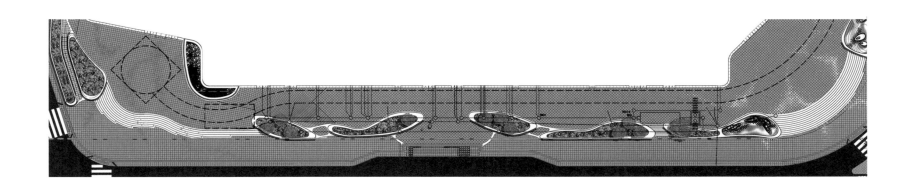

31

28 广场地面图案
29 广场台阶绿化
30 广场绿化
31 广场绿化平面图
32 与汉街连通处入口

32

NIGHTSCAPE OF PLAZA
广场夜景

汉街万达广场夜景照明的成功在于创新！国际首创
"双层媒体幕墙"，通过灯光对城市人文、艺术进行
深层挖掘与探讨，将建筑造型、表皮材质、分层成像
等设计理念完美融合，在建筑表皮整体形成既有差
异又有关联的视觉统一体，呈现出建筑的新锐风格
和华丽时尚的商业主张。汉街万达广场的夜景照明
设计，获得了2014年度第31届IALD国际照明设计大
奖，成为中国首个问鼎IALD奖项的商业地产项目，也
是中国首次有项目获得IALD建筑照明类最高级别奖
项——卓越奖。

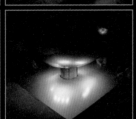

The nightscape success of Wuhan Han Street Wanda Plaza lies in its unique innovation! The internationally initiative "double media curtain wall", thoroughly studied and explored the urban humanities and arts through the design of lights, perfectly integrates the design ideas of architectural massing, surface texture, and stratified imaging. The lighting forms visual unity with both differences and correlation on the overall surface of the architecture, showing a new architectural style and a gorgeous and fashionable commercial proposition. The nightscape lighting of Han Street Wanda Plaza won the 31st IALD International Lighting Design Award in 2014. It is the first commercial real estate project in China won the IALD award and also the first to win the highest award of construction lighting-IALD Award of Excellence.

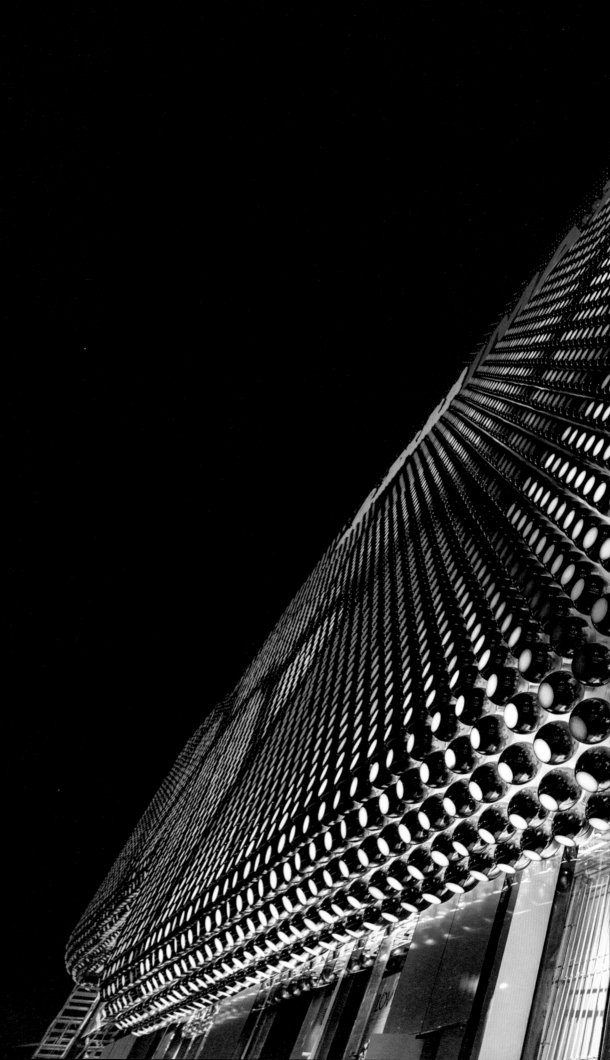

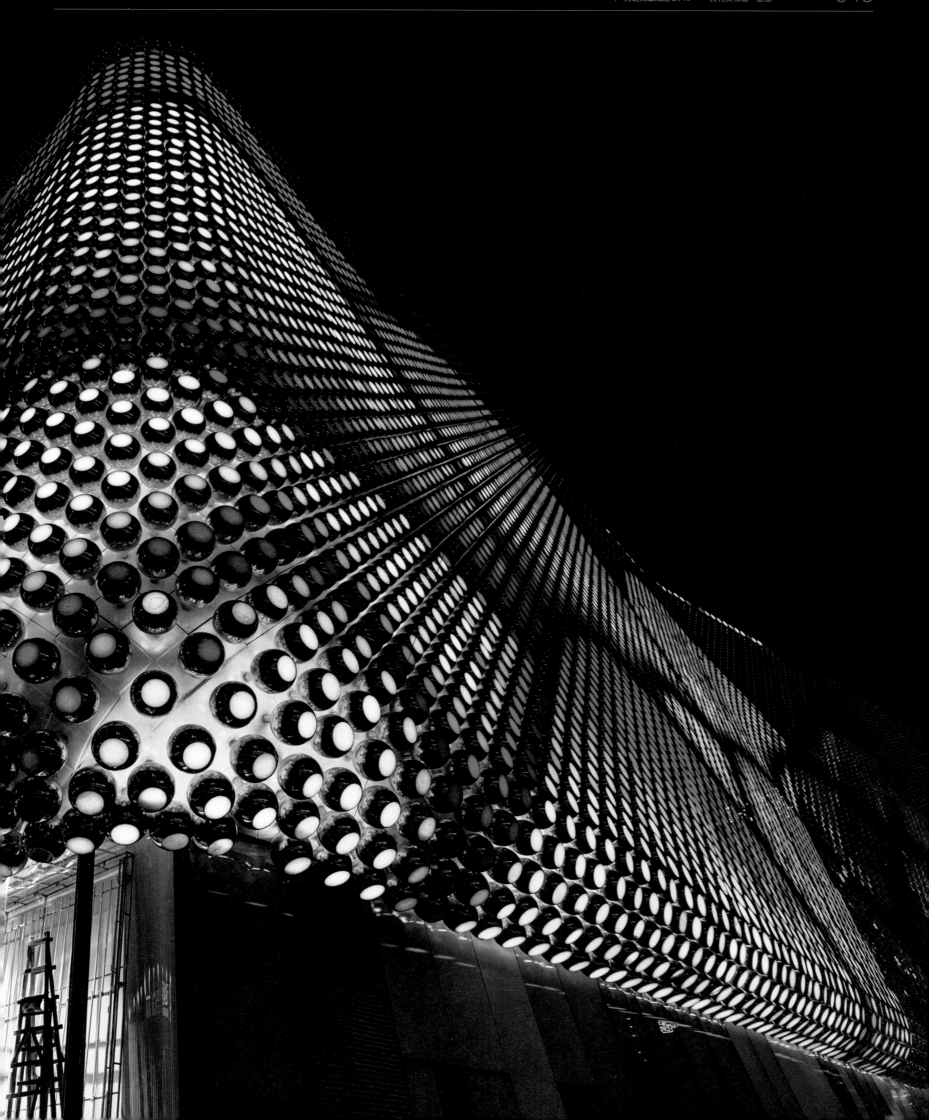

35

35 屏幕图案
36 广场夜景
37 夜景亮点及清晰成像分析

36

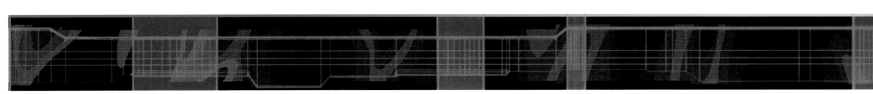

37

39

40

38 楚河对岸夜景
39 广场夜景图案一
40 广场夜景图案二

DALIAN HIGH-TECH WANDA PLAZA
大连高新万达广场

开业时间	2013 / 05 / 25
建设地点	辽宁 / 大连
占地面积	5.81 公顷
建筑面积	28.44 万平方米

OPENED ON MAY 25 / 2013
LOCATION DALIAN / LIAONING PROVINCE
LAND AREA 5.81 HECTARES
FLOOR AREA 284,400 m²

OVERVIEW OF PLAZA
广场概述

本项目规划用地5.81公顷，总建筑面积28.44万平方米，项目包括大型购物中心、高级公寓、室外步行街等，涵盖居住、购物、餐饮、休闲娱乐等功能。其中购物中心20.31万平方米，汇集万达百货、万达影城、大歌星KTV、大玩家电玩城、超市、大型酒楼、健身等多个主力店业态。

This project covers total land area of 5.81 hectares with total floor area of 284,400 m². It includes large shopping centre, luxury apartment and outdoor pedestrian street, provides housing, shopping, dining, leisure and entertainment. The shopping centre, with the area of 203,100 m², consists of Wanda Department Store, Wanda Cinema, Big Star KTV, Super Player Center, supermarket, large restaurants, gymnasium and other faciliies.

1

1 广场总平面图
2 广场鸟瞰图

FACADE OF PLAZA
广场外装

大连高新万达广场强调完整的建筑造型，具有强烈的雕塑感和造型独创性，令人激动震撼，建筑体现多肌理组合穿插的设计理念，晶莹剔透的三棱锥水晶造型间隔以简洁的铝合金条形造型。随着光影变换，呈现不同的反射，或银灰色的玻璃本身质感，或反映天光云影的变化。整体造型大气完整，赋予了整体建筑综合体强烈的意象和棱角分明的性格。

Dalian High-Tech Wanda Plaza emphasizes the consistent architectural image. With a strong sense of sculpture and originality, the architecture has a shocking design concept of multi-texture combination. The sparkling crystalline triangular pyramid blocks are spaced by concise aluminum alloy strip, which presents the changing reflections along with the lights and shadows, of the silver glass texture, and of the sky and the clouds. The overall strong integrated building image gives the architecture a sense of complexity and a distinctive character.

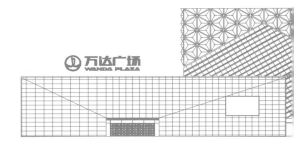

3

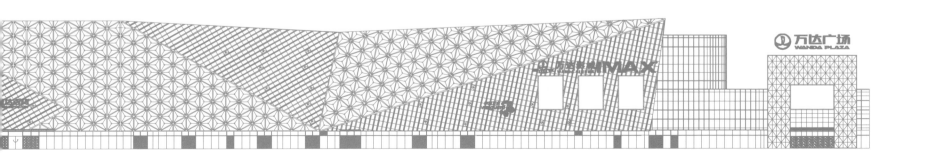

4

3　广场外立面
4　外立面图

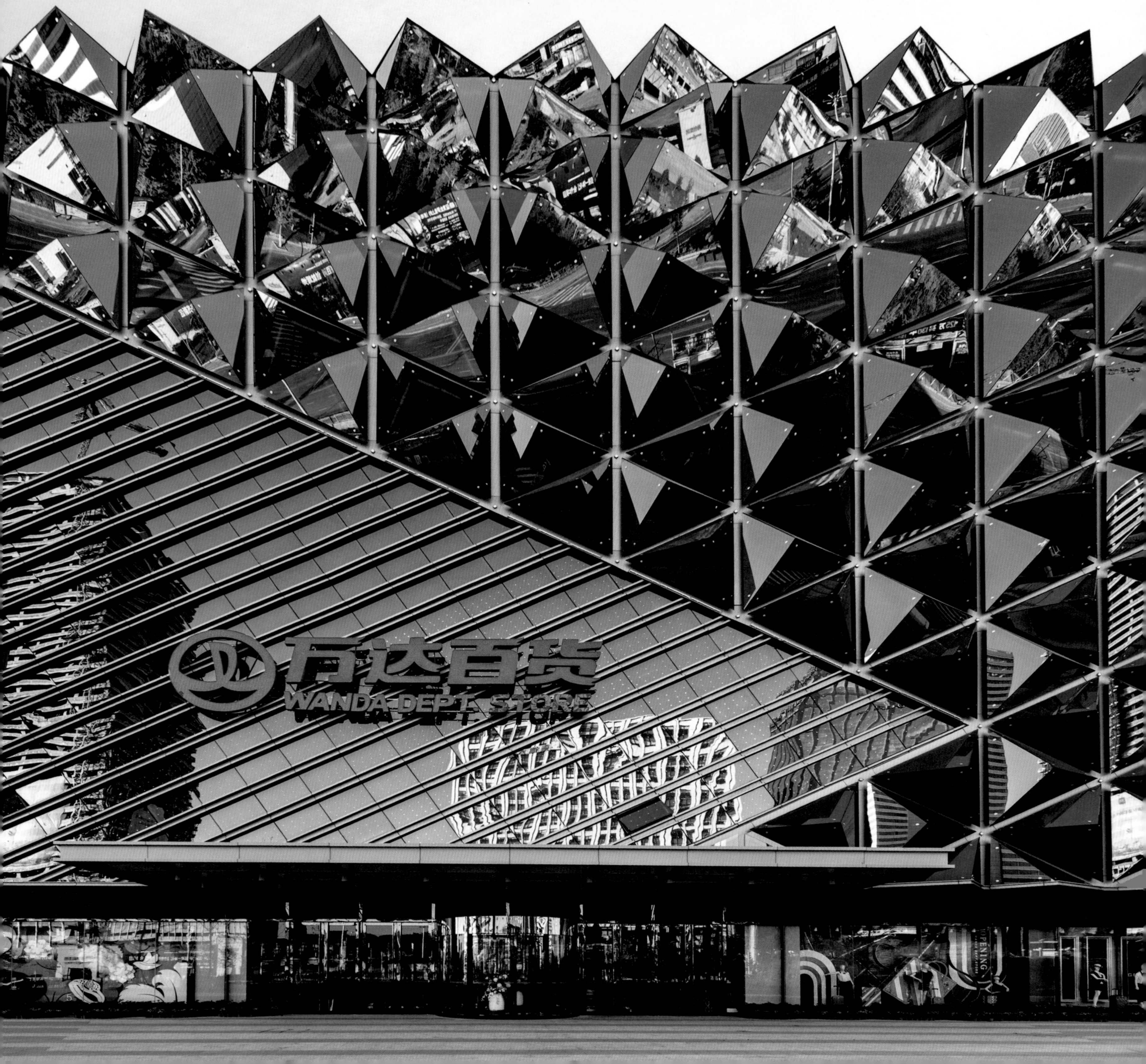
万达百货
WANDA DEPT. STORE

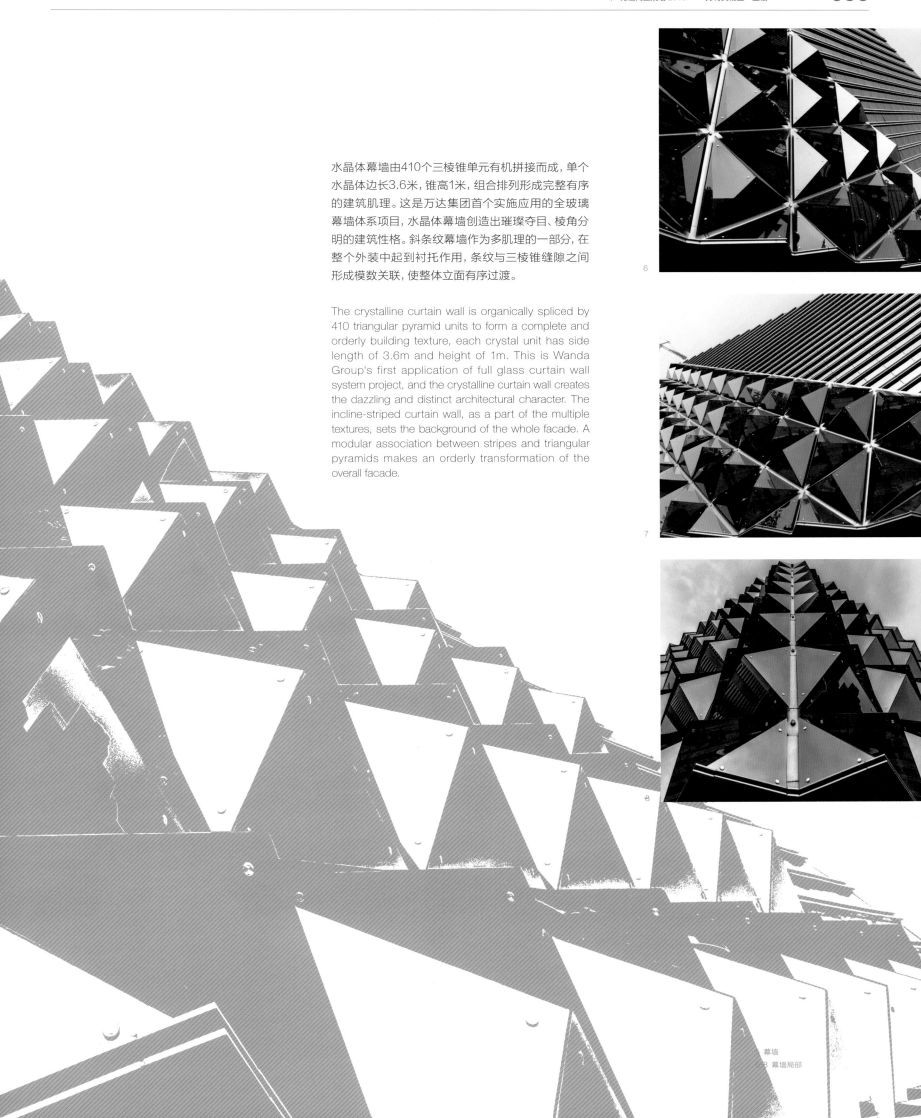

水晶体幕墙由410个三棱锥单元有机拼接而成，单个
水晶体边长3.6米，锥高1米，组合排列形成完整有序
的建筑肌理。这是万达集团首个实施应用的全玻璃
幕墙体系项目，水晶体幕墙创造出璀璨夺目、棱角分
明的建筑性格。斜条纹幕墙作为多肌理的一部分，在
整个外装中起到衬托作用，条纹与三棱锥缝隙之间
形成模数关联，使整体立面有序过渡。

The crystalline curtain wall is organically spliced by
410 triangular pyramid units to form a complete and
orderly building texture, each crystal unit has side
length of 3.6m and height of 1m. This is Wanda
Group's first application of full glass curtain wall
system project, and the crystalline curtain wall creates
the dazzling and distinct architectural character. The
incline-striped curtain wall, as a part of the multiple
textures, sets the background of the whole facade. A
modular association between stripes and triangular
pyramids makes an orderly transformation of the
overall facade.

6

7

8

幕墙
8 幕墙局部

大商业主门头设计大胆创新，不同距离出挑的金属管形成了具有张力的巨大曲面造型，曲面收缩至主入口，具有非常强的欢迎感。次门头的设计，在主体三角锥造型的基础上，选择四棱锥形的玻璃造型，既与主体呼应协调，同时富于变化；与三棱锥的尺度相比，四棱锥的尺度温和亲切，表达了次门头的细腻性格。

The design of the door frame at the main entrance is full of innovation. The metal tubes cantilevered at different distances creates a huge curve surface with great tension, which shrinks to the main entrance with a warm sense of welcome. The design of the secondary entrance adopts the rectangular glass pyramid blocks, refer to the triangular pyramid blocks at the main entrance. The rectangular glass pyramid is in harmony with the principal architectural language yet provides a different solution. Compared to the triangular pyramid, the rectangular pyramid is more gentle and sincere, which expresses the refined character of the secondary entrance.

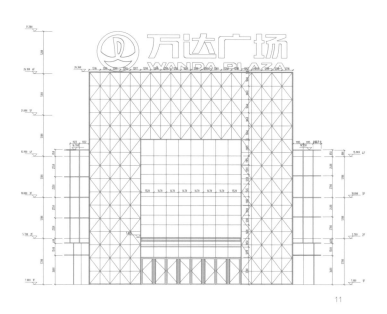

11

10　二号入口
11　门头立面图

INTERIOR OF PLAZA
广场内装

大连高新万达广场室内步行街内装设计充分发掘大连的城市文化特色，结合项目本身性质和建筑特点，重点突出室内步行街精致时尚的商业氛围，努力营造出轻松舒适、简约现代的购物环境。内装设计借助形体之间材质、肌理、色彩、灯光及疏密关系上的穿插变化，使空间简约而不单调，热闹而不喧杂。理性紧致的空间中加入空间构成，使之成为室内基本的符号元素。并结合本地特有的滨海特色及城市文化，使空间整体氛围达到和谐统一、浑然天成的艺术效果。

The interior design of Dalian High-Tech Wanda Plaza fully explores the cultural characteristics of Dalian city. Combining with the natures and characteristics of the project itself, the design highlights the elegant and fashionable atmosphere of the indoor pedestrian street, making efforts to create a relaxing, comfortable and chic shopping environment. By the interludes and changes of the materials, textures, colors, lights and the relations between the forms, the interior design makes the space simple but not drab, lively but not noisy. Adding spatial constitution to the rational and efficient space makes it one of the basic indoor elements; the integration of the local unique coastal features and the city culture make the overall space atmosphere in an artistically harmony.

12 室内步行街

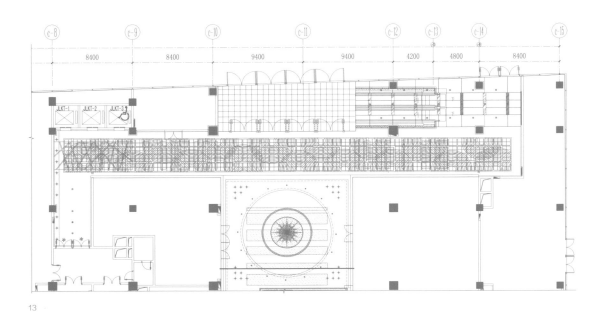

13

一号门主入口设计考虑各部分的功能需求，在狭长的横向通道天花设置一个LED天幕，由LED灯组成圆形图案，按照预计的频率开闭，最后形成了一个波光粼粼的水面效果，行走下面有巨大的震撼感，犹如行走在海底世界仰望海面，突出了以海洋为主题的购物环境。小厅的顶面为了增加引导性，设置了圆形的LED灯池，内置钻石型的LED灯珠，通过镜面的折射，使圆形的灯盒更像夜晚的星空，具有海天一色的意境。

Considering the function requirements of different part, an LED screen is designed on the long lateral ceiling at the main entrance of Gate 1. LED lamps compose a circular pattern, with an expected frequency of on and off to form the effect of glittering water. When walking under the screen, there is a huge sense of shock like walking under the sea looking up to the sea level. It highlights the ocean theme of the shopping environment. In order to enhance the guidance, a round LED light pool is set on the top surface of the small hall. The built-in diamond shaped LED lamps, by the mirror refraction, decorate the round light pool like the sky of a starry night, blending the horizon between the sea and the sky.

14

13 入口天花平面图
14 入口天花
15 入口门厅

在椭圆中庭，围绕观光电梯周围的侧裙板采用方格纹理的发光玻璃，显露出晶莹剔透，美轮美奂的艺术效果；大面积侧裙板通过体块的穿插，以及漂浮设计手法，营造出轻盈灵动的空间特色；地面若隐若现的方格纹理拼花打破了地面的呆板沉闷，图案犹如花格地毯，尽显华贵典雅气质。

In the elliptical atrium, luminescence glass with grid pattern is used for the side skirt of the sightseeing elevators to show the translucent and magnificent artistic effects. The intercross of the massive side skirts and the floating design create the spatial characteristics of lightness and vividness. The floor parquetry with indistinct grid pattern breaks the ponderousness, and the pattern pavement, which like a beautiful carpet, shows the luxurious and elegant temperament.

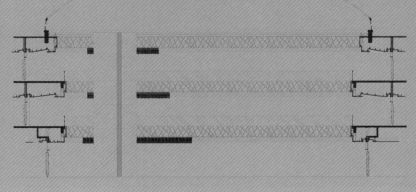

16
17

16 椭圆中庭
17 椭圆中庭剖面图

圆中庭采用了体块穿插的手法，两种颜色、质地强烈对比，直线体块在空间中穿插，以及灯光营造的错位效果等，形成了交错有序的视觉效果，也使圆中庭的视觉进深感加强，从而产生出较大的空间感。圆中庭侧裙板分为两种材质与形态，分别赋予灰金色和白色，同时注重细节处理，如相同模数设计、表面质感肌理、日光和灯光对于材质表面形成的光效影响、两种材料的空间比例关系等，均经过反复琢磨，形成极具个性的中庭空间。另一处特色效果体现在观光电梯转折面，透亮的方格元素，辅以背光处理，更显品质，令圆中庭空间时时流露出简约时尚、精致尊贵的艺术效果。

The circular atrium adopts the interweaving technique between blocks, and the strong contrast of two different colors and textures, the linear blocks interspersed in the space, and the dislocation effects created by the lighting form the staggered but orderly visual effects, all these manners strengthen the visual sensation of spatial depth in the circular atrium, and enhance the sense of the grand space. The side skirts used in the circular atrium have two types of materials and shapes, given the color of grey gold and white. Meanwhile the details have been carefully studied. The same modular design, surface texture, light effects of the material surface by the sunlight and artificial lights, and the spatial proportions of two materials are pondered over and over again to finalize the distinct atrium space. Another characteristic effect is embodied on the flexure surface of the sightseeing elevators. The transparent and shining grid element with the backlighting, which shows superior quality, reveals the chic, exquisite and noble effects of the circular atrium all the time.

18

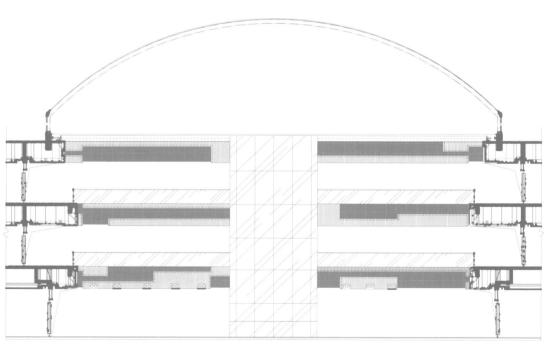

19

18 圆中庭
19 圆中庭剖面图
20 圆中庭采光顶

21

在长街空间中, 采用平铺直叙的表现手法, 强调侧裙
板的肌理和色彩的搭配, 侧裙板GRG材质细微的凹
凸和精致的型材收口等细节处理, 令空间极具细节
品质; 同时, 在室内空间中首次使用"模数化"设计,
将外露的设备与顶面造型、侧裙排板、地面拼花等按
照一定的比例模数有机排列, 使长街达到井然有序、
透视感强烈的视觉效果。

On the pedestrian street, a simple expression
emphasizes the texture of the side skirt panels and the
color collocation. The subtle concaves and convexes
of GRG side skirt panels and the delicate cuffs profile
bring the space with quality in every detail. Meanwhile,
the "modular" design method is applied to the interior
space for the first time. The organic arrangement of
the exposed equipments, ceiling shape, side panels
layout and floor pavement pattern are all set in certain
proportion, in order to create the orderly visual effect with
a strong sense of perspective on the pedestrian street.

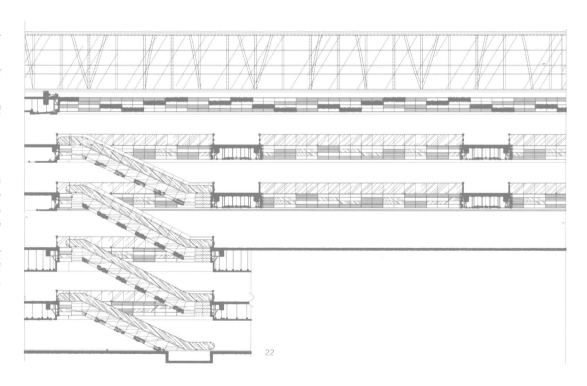

22

21 室内步行街
22 室内步行街剖面图
23 连桥
24 自动扶梯
25 室内步行街

23

24

LANDSCAPE OF PLAZA
广场景观

鱼盘雕塑设计紧密围绕"海洋"主题，立意鲜明，造型新颖别致。鱼盘雕塑与人形雕塑组合摆放，寓意人与海洋和谐共处。鱼盘与人形雕塑完美反射出1号门门头绚烂变幻的光影。人形雕塑尺度亲人，成为市民最爱合影留念互动嬉戏的地方。绽放雕塑高度8米，材质为拉丝不锈钢，立于主广场水池中央，形如四朵即将绽放的花苞，又似喷涌飞溅的水柱，预示着大连高新万达更加绚丽的明天。沉稳黑色的水池、活泼跳动的涌泉、炫目银白的雕塑犹如一段和谐的乐章、一曲绝妙的赞歌。

Closely around the theme of "Ocean", the design of the "fish plate" sculpture has a purposive conception and a novel shape. The placement of the fish plate sculpture combined with human-shaped sculpture symbolizes the harmonious coexistence of human and the ocean. The fish plate and the human-shaped sculptures perfectly reflect the gorgeous change of lights and shadows on the door frame of Gate 1. The human-shaped sculpture is friendly in sense and becomes a residents' favorite place for taking photos and interactive leisure. The blossom sculpture, with 8 meter height with brushed stainless steel, stands in the fountain of the main square like four flower buds which are about to bloom, or like the splashing water spray, to symbolize a more brilliant future of Dalian High-Tech Wanda Plaza. The black pool, lively fountain and the dazzling silver white sculpture compose a harmonious musical movement, or a wonderful hymn.

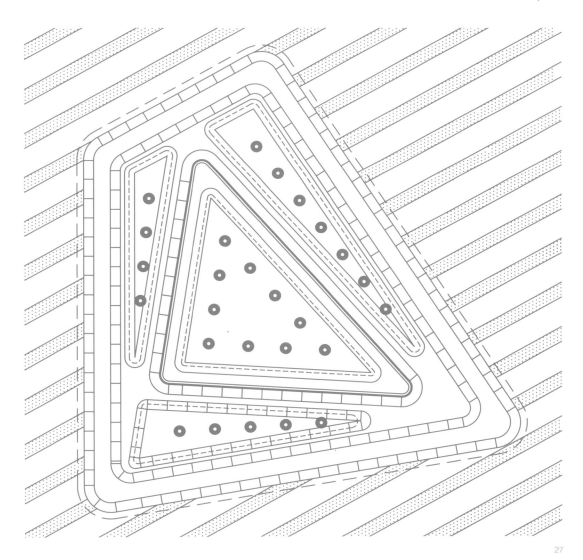

27

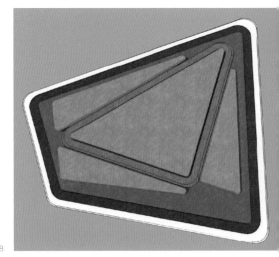

28

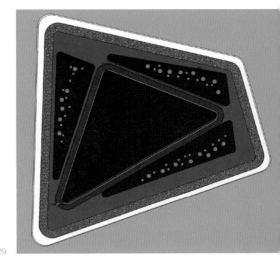

29

26 广场水景
27 广场水景平面图
28 水景形式——有水平面
29 水景形式——无水平面

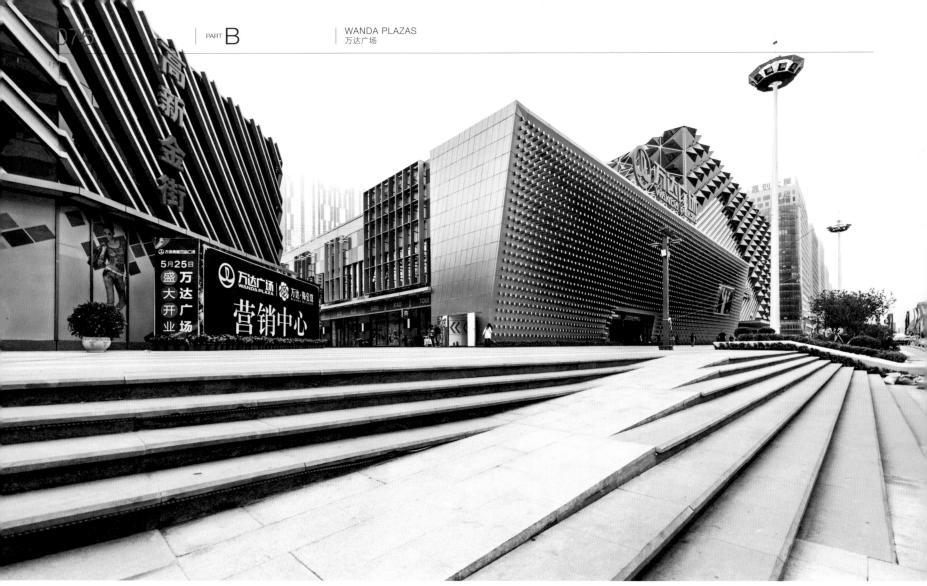

30

31

32

33

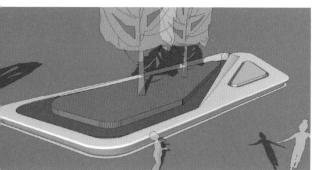

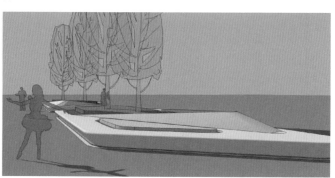

30 广场台阶
31 广场沿街花坛
32 沿街花坛效果图一
33 沿街花坛效果图二
34 广场雕塑

NIGHTSCAPE OF PLAZA
广场夜景

大连高新万达广场在照明设计之初，就确立了体现万达广场国际化品牌形象的目标，明确了将其打造成国内商业建筑照明标杆的定位。建筑立面被流动的光带有序地分割成多个切面，切面顶部流光闪动，仿佛钻石的顶点。每个切面颜色缓慢变化，三棱锥水晶体区域则通过七彩的灯光形成钻石般闪耀的效果，并有流光划过。通过"三棱锥"与斜条纹的灯光组合，形成了体现大连独特人文气息的，以"浪漫、时尚、海洋、历史、高科、人文、动感、未来"为主题的8个章节的视频动画，用灯光诠释了一幅优美、靓丽的动画场景，赋予了大连高新万达广场"硅谷海岸、浪漫之都"的象征意义。

At the very beginning of the lighting design, Dalian High-Tech Wanda Plaza established the objective to embody the international brand image of Wanda plaza and cleared its core design concept to build the lighting benchmark of the national commercial buildings. The building facade is orderly divided into multiple sections by the floating light with the top of the section flashing like a diamond, and the color of each section changes slowly. The shining diamond effect of the triangular pyramid area is formed by the colorful lights. The combination of the "triangular pyramids" and the inclined light stripes create the video animation which include 8 chapters with the theme of Romance, Fashion, Ocean, History, High-Tech, Humanity, Dynamic and Future respectively and express the unique humanistic atmosphere of Dalian. A beautiful animation scene interpreted by the lights gives Dalian High-Tech Wanda Plaza the symbolic significance of "coast of Silicon Valley, city of romance".

35 夜景鸟瞰图

36

37

38

39

40

36 一号入口夜景
37 二号入口夜景
38 二号门头夜景
39 外立面夜景一
40 外立面夜景二

OUTDOOR PEDESTRIAN STREET
室外步行街

室外步行街是整个商业综合体的重要组成部分，设计过程中，采用玻璃与不同色彩的金属幕墙的对比，同时强调和大商业建筑的整体性，底层的设计除了通透的橱窗，同时选用了间隔色彩的暖色铝板幕墙，温暖舒适。美陈方面，更多地以雨篷色彩和丰富的侧招造型为主，并利用竖向金属挑板和构件的色彩变化突出室外步行街的商业氛围。

The outdoor pedestrian street is one of the important parts of this commercial complex. During the design procedure, glass is in troduced to contrast with metal curtain walls of different colors, which emphasize the integrality of the commercial building. The design of the bottom floor adopts transparent windows and aluminum curtain walls with interval of warm colors which is warm and comfortable. Speaking of the decoration, it relies on the color of the canopy and the rich shaped side signage. The changing color of the vertical cantilevered metal panels and the components highlight the commercial atmosphere of the outdoor pedestrian street.

42

41 商业街夜景
42 商业街入口
43 商业街外立面

43

YIXING
WANDA PLAZA
宜兴万达广场

开业时间 2013 / 05 / 31
建设地点 江苏 / 宜兴
占地面积 12.08 公顷
建筑面积 52.08 万平方米

OPENED ON MAY 31 / 2013
LOCATION YIXING / JIANGSU PROVINCE
LAND AREA 12.08 HECTARES
FLOOR AREA 520,800 m²

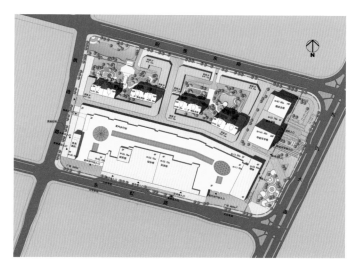

OVERVIEW OF PLAZA
广场概述

宜兴万达广场项目总占地12.08公顷，总建筑面积52.08万平方米，项目包括购物中心、室外步行街、酒店、甲级写字楼、豪宅等，是集居住、购物、餐饮、休闲娱乐、办公等多种功能于一体的复合性大型城市综合体。购物中心21.51万平方米，汇集万达百货、万达影城、大歌星KTV、大玩家电玩城、精品超市、大型酒楼等多个主力业态。

设计中，更多体现对环境的尊重，从商业、景观、城市设计等多角度阐释万达广场对人和城市的影响。从规划入手，细致考虑宜兴的城市肌理和文脉，将大商业和酒店、写字楼等重要物业沿主要道路展开。室外步行街形成项目的外部轴线，衔接各部分。将西北角水系景观引入地块，结合设计高档住宅，充分利用土地，力求布局合理，空间紧凑，环境宜人。沿东虹东路和东氿大道布置商务酒店、购物中心、五星级酒店及甲级写字楼。

This project covers total land area of 12.08 hectares with total floor area of 520,800 m². Including shopping centre, outdoor pedestrian street, hotel, Grade "A" office building and mansions, it is a large-scale composite city complex which covers the functions of housing, shopping, dining, leisure, entertainment and office. The shopping centre, with the area of 215,100 m², consists of Wanda Department Store, Wanda Cinema, Big Star KTV, Super Player Center, boutique supermarket, large restaurants and other activities.

The design respects the context and interprets how a commercial complex influences people and the city in terms of commercial, landscape and urban design. Starting from the master planning, the design carefully considered Yixing's texture and context, and develops the major commercial, hotel, office building and other important properties along the main road. The outdoor pedestrian street, as the outer axis, connects each part of the project. The water feature is introduced and combined with the design of high-end residential, which strives fully use the land for rational layout, compact layout and pleasant environment. Business hotel, shopping centre, 5-star hotel and Grade "A" office building are arranged along Donghong East Road and Donggui Road.

1 广场鸟瞰图
2 广场总平面图

FACADE OF PLAZA
广场外装

宜兴万达广场大商业面宽较大，因此设计中创造性地模糊内部功能边界，通过连贯延续的水平线条、丰富的肌理变化，构成建筑的主立面，将广告店招等商业元素穿插其中，形成整体且丰富的立面造型，简洁且具有动感和冲击力，进而最大限度地发挥基地自身的特质，营造了大气、连贯而不单调的城市商业界面。

主入口门高19.3米，入口色彩亮丽，优雅大方，同时又突显其标志性与重要性，是本项目的创新亮点之一。购物中心主入口，以红色的彩釉玻璃为门头的主色，辅以精心设计的中国结底纹，内部装饰以亚克力灯管组装成中国结图案，形成"大红门里的吉祥结"，充满着中国文化底蕴。

Due to a large commercial range, the design of Yixing Wanda Plaza creatively blur the boundary of each internal function. The consistent horizontal lines and the various textures form the facade of the building. Advertising signage and other commercial elements are interspersed among them to form a complete and rich look full of dynamic and impact, which therefore maximums the own characteristics of the project and creates the grand and coherent urban commercial interface.

The main entrance, with the height of 19.3m, is colorful and elegant and highlights its landmark and importance, so it could be considered as one of the innovation spots of the project. The main entrance of the shopping centre is glazed by red glass as the main color with the elaborate shape of Chinese knot shading. Acrylic tubes are assembled into Chinese knot pattern for internal decoration, which brings "auspicious knot in the big red gate" with the full Chinese culture.

3 广场一号入口立面图
4 广场一号入口

幕墙质感独特，色泽持久，而且外观形状可以多样化。宜兴万达广场中采用大面积的铝单板幕墙、穿孔铝板幕墙、网格铝板幕墙，多种形式相互穿插，并与玻璃幕墙完美地结合，强调了大商业立面的水平线条和体量感，形成整体且丰富的立面造型。

The curtain wall has the unique texture, durable color and varied shape. The curtain wall of Yixing Wanda Plaza used massive single-plate aluminum panels, perforated aluminum panels and grid-patterned aluminum panels. The mutual interpenetration of various forms combined perfectly to the glass curtain wall, and emphasizes horizontal lines and dimension sense that enriches the shape of the facade.

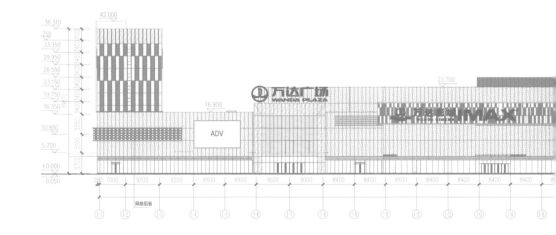

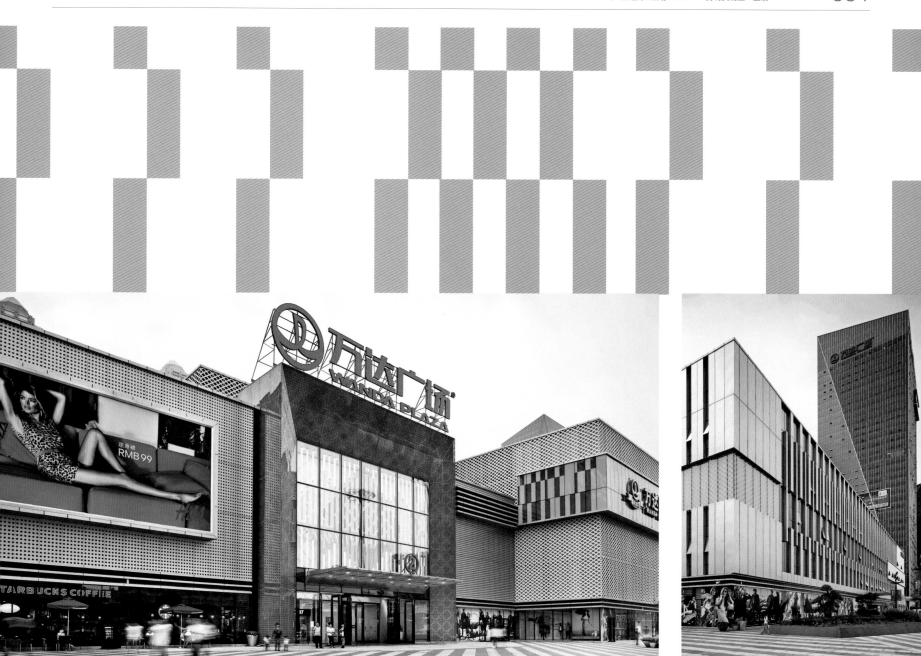

6

7

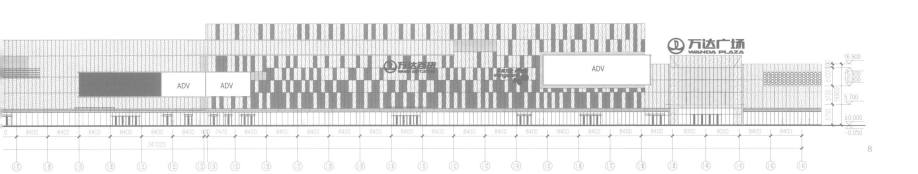

8

5　广场百货外立面
6　广场二号入口
7　广场外立面
8　广场立面图

INTERIOR OF PLAZA
广场内装

设计之初，对于江南的美，在诗里陶醉。多少次幻想，秦淮河上"烟笼寒水月笼纱"的如梦如幻。多少次向往，江南水乡里"小桥流水人家"的古朴与安闲，在这般幻梦中找寻设计理念，几经易稿，多次的方案讨论，多次推翻了已有的设计初稿，从整体江南意境入手。以江南美景为意，鱼水之乡为形，意与形相结合，用中式简约的手法，表现婉约之美。另一方面，在细节和气质的把握上，泼一纸重墨，用设计感强烈的现代小品和工艺品点缀商业空间的各个角落，并结合精湛的施工工艺和细致入微的人文关怀，向顾客传递空间的品质和亲和力。力图打造一个充满时尚感、尊贵感、有内涵的现代娱乐购物之都。

At the beginning of the design, the beauty of Jiangnan is only revealed in the poems. How many times have we imagined the cold water and white sand shrouded in the moonlight and the light smoke above Qinhuai River? How many times have we dreamt the simple and leisure lifestyle with bridge, stream and house in the region of Jiangnan? The idea of the design is conceived from such dreams. After several theme discussions and a few versions exploriation, the design starts to form the overall Jiangnan conception and combines the meaning of Jiangnan beauty and the form of water town to express the graceful beauty in a simple Chinese style. On the other hand, the design pays much attention to the details and the quality. A series of modern artworks and crafts are decorated in the corners of the commercial space. The superb workmanship and the meticulous humanistic care communicate the quality and affinity of the space to customers, trying to build a modern entertainment and shopping centre full of fashion sense, dignity and connotation.

9 椭圆中庭采光顶
10 椭圆中庭

将外立面条形拼贴的图案引入到室内，与建筑外立面互相呼应，每一个小细节，都如一道独特的风景，和谐统一，通过不同材质拼贴组合，将作为主入口功能的天花，不容修饰，不容雕刻，贯穿整个长街，使整个动线上的节奏高低起伏。天花造型犹如四扇开启的门，迎接四方宾客的到来。

二道门区域从江南花格窗中提取设计元素，应用到天花、墙面石材造型上。为烘托商业氛围，在每个凸出的石材上都暗藏了LED灯，使整个空间充满江南水乡的独到风景，再配合外立面的彩色发光中国结，使江南style发挥到极致。顾客在进入购物中心之前，犹如打开一扇江南民居的窗。

The strip pattern collaged for the facade is introduced to the interior for the mutual echo. Every little detail, as unique scenery, realizes harmony and unity. The ceiling at the main entrance is collaged by the combination of different materials through the entire street without any more modification or carving, which increase the rhythm of the dynamic. The shape of the ceiling, like four open doors, welcomes the guests from all directions.

In the region of the second entrance, the ceiling design is conceived from Jiangnan lattice windows. In order to strengthen the commercial atmosphere, an LED lamp is hidden in each bulged wall stone, creates the unique Jiangnan scenery in the whole space. The integration with the light shining Chinese knot on the façade, it plays Jiangnan style to extreme. When Customer entering the shopping centre, the impression is like open the window of a Jiangnan house.

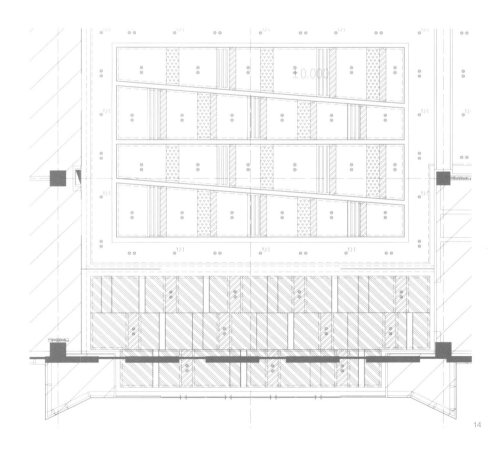

14

11 入口天花
12 入口门厅
13 入口
14 天花平面图

椭圆中庭同样也是从花格窗中提取元素,运用在设计当中。在整体空间中,造型组成的序列,增强了空间的动势。犹如远处青山围绕的一缕缕白雾,整体调子简洁素雅,设计运用多种材质对比,提升了整个购物中心的档次。

The elliptical atrium is also conceived from the element of Jiangnan lattice windows. In the whole space, the sequence of the shapes enhances the dynamic state. The overall tone is concise, simple but elegant, like white mist around a green mountain in distance. A variety of contrast by materials have improve the grade of the whole shopping centre.

15

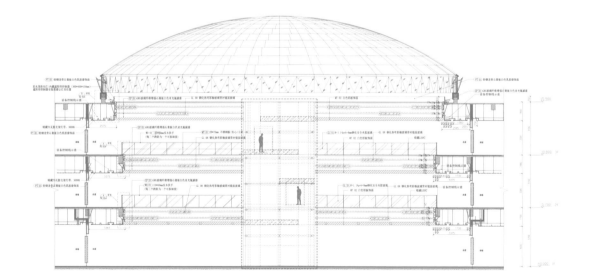

16

15 椭圆中庭
16 椭圆中庭剖面图
17 椭圆中庭

说到江南，总会想到江南雨水。那包含诗情的画卷中总少不了细细的雨丝弥漫于小桥流水人家中，增添些淡淡的愁意。圆中庭的设计是把鱼和水作为设计元素，将其交融，形成图案。观光电梯立面设计强化了江南这一元素在整个空间中所营造的氛围，借助形体之间、灯光及疏密关系上的穿插变化，使空间简约而不单调，丰富而不凌乱，热闹非凡而又不喧杂，仿古而不复古。使室内室外空间整体氛围达到和谐统一，浑然天成，显示江南独有的韵味。

Speaking of Jiangnan, people would be reminded of the rain. The poetic pictures always contain the melancholy of light rain permeating bridges, streams and houses, The circular atrium takes fish and water as the design elements and blends them to create the pattern. The elevation design of the sightseeing elevators emphasizes the atmosphere which Jiangnan elements create in the whole space. By the interludes and changes of the forms, lights and their density proportions, the interior design makes the space simple but not drab, rich but not messy, lively but not noisy, dassical but not obsoleted. The harmony and unity of the interior and exterior expresses the unique charm of Jiangnan.

19

20

21

18 圆中庭
19 圆中庭天窗
20 圆中庭侧帮
21 圆中庭剖面图

长街的设计同样也是延续花格窗元素,侧裙板由铝板及GRG两种不同的材质组成不规律的格子,整体设计高贵清雅。连接长街的廊桥采用同样的设计手法,将镜面和彩釉玻璃相结合,削弱廊桥对整体空间的影响,增加通透性。使整个空间如诗如画般的端庄,有着娇柔淡雅的气质,令人思绪万千、回味无穷。

The design of the straight gallery also follows the lattice window element, and the side plates consist of aluminum and GRG materials to form irregular grids, making the whole design appear elegant and graceful. The bridge connecting the straight gallery applies the same design method to combine the mirror and ceramic glaze, undermines influence of the bridge on the overall space, and consolidates the sense of permeability. Through this design, the whole space appears poetically delicate and elegant, attracting people to long indulge in this space.

22

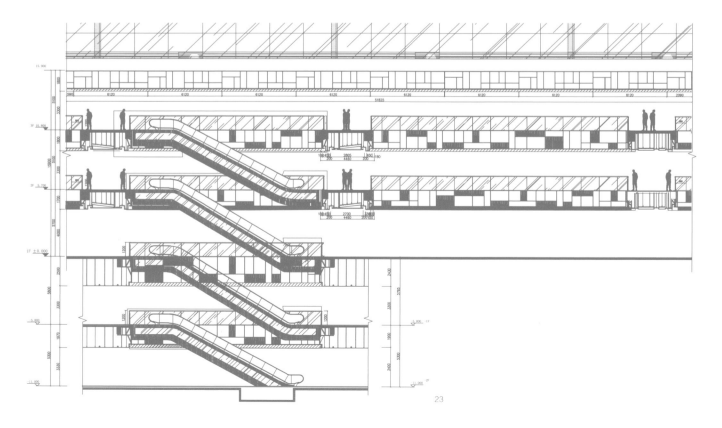

23

24

22 自动扶梯
23 室内步行街剖面图
24 室内步行街

ONE STORE, ONE STYLE
一店一色

在室内商业街的设计中，为了更好体现整个购物中心的档次和"时尚生活中心"的核心定位，要求各商户发挥各自品牌的独有特色，把握和放大自身品牌风格优势，将"一店一色"的装修特色发挥得淋漓尽致。这种各具创意风格的装修特色令人耳目一新。

In order to better reflect the grade of the whole shopping centre and its core positioning of "fashion and living centre", the indoor pedestrian street design requires that each merchant has on unique characteristics of its own brand and amplifies the brand style and advantages to vivid the decoration character of "one Store, one style". This unique and creative character of decoration creates a refreshing impression.

25a

25b

25 特色店面

27

28

29

LANDSCAPE OF PLAZA
广场景观

宜兴万达广场的景观设计与建筑相呼应，从建筑外立面汲取设计灵感。按照以人为本、服务商业的设计原则，不断追求创新的设计精神，采用现代极简主义的设计手法，创作出了引导性极强的条带格式铺装；时尚简洁的立体花池；色彩丰富、形象鲜活的雕塑；精致洗练的种植设计及卓有情趣的设施小品。这些丰富的设计手法的运用，营造出具有独特的可识别性和热烈的商业氛围。

The landscape design echoes the architecture and is conceived from the facade. According to the design principle of people-oriented and commercial service, the design techniques of modern minimalist are adopted to create striped pavement with strong guidance, fashionable and concise 3D flower bed, colorful and vivid sculpture, exquisite planting design as well as interesting facilities. The rich design manners bring the project with unique identity and warm commercial atmosphere.

26 广场景观
27 广场绿化带
28 广场花坛
29 广场雕塑

NIGHTSCAPE OF PLAZA
广场夜景

宜兴万达广场的夜景照明设计，综合考虑宜兴万达广场地处东氿新城的正中心，具有优越的地理位置。对宜兴的历史人文、名胜山水等城市元素进行提炼，设计风格体现出宏伟、大气、璀璨、靓丽高品位的效果。通过不同灯光对室外步行街、豪宅建筑细部加以表现，使万达广场不同建筑展现精致典雅的夜间面貌。营造舒适宜人的夜间光环境，提升万达广场的层次品位，让夜幕当中的万达广场成为城市的亮点。凸显"一座万达广场、一个城市中心"的口号，体现了"科技历史文化、生态自然人文"的设计思路。

The nightscape design takes into account the location advantage of Yixing Wanda Plaza right at the centre of Dongjiu New Town. The design style is refined from Yixing's history and culture, scenic landscape and other elements of the city, and reflects magnificent, grand, beautiful, bright and superior premium effects. Different lights focus on the outdoor pedestrian street and the architectural details of the mansion to express a delicate and elegant night scene. The comfortable and pleasant nightscape improves the level of Wanda Plaza, highlighting the plaza in the night city. The design emphasizes the slogan "One Wanda Plaza, One City Centre", and reflects the idea of "technology, history and culture, ecology, nature and humanity".

30 广场外立面夜景

32

大商业建筑立面大面积的穿孔板和实体幕墙铝板的运用，为夜景照明带来了良好的展示空间。在建筑穿孔板立面采用LED像素点背投的照明方式形成影像，巧妙展现了见光不见灯的照明设计理念。大商业建筑实体幕墙采用内嵌LED面板灯。在视觉手法上，通过一系列巧妙设置的灯光动态元素，与呈现的画面对接，浑然一体；同时，设置绿色元素作为视觉引领，在潜移默化间传达"绿色环保"照明理念的同时，给人以积极向上、生机勃勃之感，强调商业的热烈氛围。

The use of massive perforated plates and solid aluminum plates on the facade provide good display space for lighting. LED pixel projection is adapted to the perforated facade for image formation, which cleverly shows the design concept "visible light, invisible light fixture". The solid curtain wall of the shopping centre adopts embedded LED panel lights. In terms of visual technique, a series of dynamic light elements are skillfully arranged to perfectly match the images displayed. At the same time, green element is used as a visual guide which seeks to subtly communicate the green lighting concept and create the positive, lively and warm commercial atmosphere.

31 大商业夜景
32 立面图

31

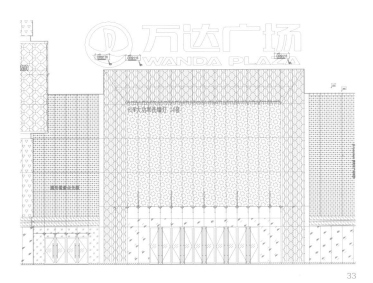

33

34

35

33 门头立面图
34 一号入口
35 二号入口

OUTDOOR PEDESTRIAN STREET
室外步行街

宜兴地处天目山余脉，蕴藏着丰富的陶土矿源，制陶史迄今已有七千余年，其中紫砂陶土更是宜兴特有的，紫砂壶也是发源于宜兴。室外步行街景观设计以宜兴独特的"紫砂文化"作为设计主题，将传统紫砂壶元素抽象成色彩丰富的雕塑小品，贯穿始终。

Located on the edge of Tianmu Mountain, Yixing has the rich source of clay minerals and the pottery history of more than seven thousand years so far. Purple clay is especially endemic in Yixing as it is the origin of purple clay pot. The outdoor pedestrian street landscape takes the unique purple clay culture of Yixing as the design theme. The traditional purple clay pot elements are abstracted into colorful sculptures throughout the street.

37

36

38

39

36 商铺及室外步行街
37 景观小品
38 室外步行街外立面
39 室外步行街特写

FACADE OF HOTEL
酒店外装

相似而不重复的形体组合，以简练的材料语言塑造
挺拔效果，微妙的折面与折线使建筑变得典雅别
致。豪华且高品质的功能需求要求建筑立面丰富且
典雅。通过层间墙的变化，形成编织感的立面造型，
尽显其高端品质。

The combination of similar but not duplicated forms
creates the forceful effect with concise material
language. The delicate sections and lines make the
building elegant and chic. The function demands for
luxury and high quality require rich and elegant facade
of the building. The change of facade by floor forms
the woven sense and show the luxury quality on
the facade.

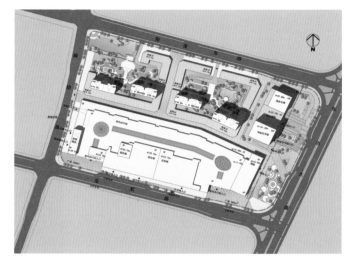

40

41

40 总平面图
41 酒店入口
42 酒店及甲级写字楼外立面

43

44

45

LANDSCAPE OF HOTEL
酒店景观

宜兴艾美酒店是宜兴地标级的酒店建筑，景观设计遵循高贵、典雅、大气的设计风格，打造五星级酒店应有的景观气质——"低调的奢华"。景观与建筑风格相结合采用现代简洁的设计手法，在酒店主入口设计了大体量的水景观，采用层跌式、对称的形式，突显酒店前场大气、恢宏的气势。酒店主入口处几株特大乔木、点景造型树及层次丰富的植物配置，两侧及水池上的金色特色景观灯，使酒店前场气势磅礴、高贵典雅。

酒店花园采用自然式的水景，黄石驳岸，舒适的木平台，精致的植物造景营造了静谧舒适的室外用餐和休息空间。景墙的设计既满足了遮挡车库的功能要求，还满足了全日餐厅观景的效果，设计中发掘宜兴传统文化元素，以当地历史文化名人苏东坡的词、画为主题设计景墙，赋予了景墙艺术性和文化韵味。酒店植物造景颇为考究，以地形设计为依托，从空间和功能的需求考虑植物景观的设计手法。开敞空间，疏可跑马，干净的草坪点缀特大乔木及特型植物，如酒店前场；私密空间，密不透风，多层次、多色彩、多形态的植物搭配，像风景画一般映入眼帘，如酒店花园。不同空间采用不同的配置手法，营造优美、色彩丰富的植物空间。

As one of the landmark buildings of Yixing, Wanda Meridien Yixing landscape design follows the noble, elegant and grand style, and makes landscape effect temperament that a 5-star hotel deserves-humble and luxury. The landscape is integrated with the architectural style by modern and simple design manner. A large volume of water landscape is design at the hotel main entrance, which, in its layered and symmetrical form highlights the grandness and magnificent atmosphere in front of the hotel. A few huge arbors, decorative trees and plant layout with rich layers are arranged at the main entrance. Golden landscape lamps on both side and above the pool present the majestic and noble elegance of the hotel.

The hotel garden adopts natural waterscape with yellow stone banks. The intimating wooden platform with refined planters creates a peaceful and cozy space for outdoor dining and relaxing. The landscape wall not only block-off the parking space, but also provides the focus element for the all-day dining restaurant. Excavating the traditional culture of Yixing, the landscape wall is designed on the basis of the historical and cultural celebrity Su Dongpo's poem and paintings, which gives the flavor of art and culture. The plants are very exquisite. Based on the terrain, the design of the plants considered both spatial and functional requirements. For example, in the hotel front court, the open and large area is laid with clean grass dotted by huge arbors and decorative plants. And for the hotel garden, the private space is domain by the plant collocation of multiple layers, colors and forms to create a landscape painting. Different spaces with different arrangements, elegant and colorful plant space are therefore created.

43 酒店水景
44 酒店景观小品
45 喷泉灯柱造型
46 绿化带

NIGHTSCAPE OF HOTEL
酒店夜景

高大建筑的灯光最注重的是气氛的烘托,面对设计精美的建筑,它高低错落围合模式,气势非凡,通过灯光的营造为建筑夜晚增辉,并能够充分表现出写字楼的典雅和酒店的温馨,从而实现照明真正的意义……夜景充满个性、特色、魅力,让高耸的建筑为城市整体夜景画上点睛之笔,成为城市的地标式建筑夜景。以LED点、线的结合自然地镶嵌在建筑表面(线一水、点一水滴),静止的画面寓意着"上善若水",节日里在夜晚中形成一副大的影视背景墙,以优美大气的画面,辅以简洁流畅的剪辑,再加上轻松愉悦的背景音乐,给人们带来强大视觉吸引的同时,留给大家充分空间来想象、回忆、感动。

For high-rise building, the focus of its facade during nights is the atmosphere set off by its exterior lights. Installed on delicately designed buildings, the lights are well arranged in a staggered layout to create extraordinary effect for the building in night, fully presenting the elegance of the office building and warmness of the hotel, and realizing the practical value of lighting. Full of characteristic and charm, the nightscape seems a finishing touch for the high building in the background of city's night, and makes the building a landmark of the city. The natural combination of spots and lines of LEDs are presented on the building's surface (lines: water; spots: drops), and the static image implies that water is the source of life. It becomes a huge film wall in festival nights; and with such grand and beautiful image, concise and smooth cutting, as well as relaxing music, the nightscape attracts people with powerful visual image and brings a large space of imagination, memory and sensation.

SHENYANG OLYMPIC WANDA PLAZA
沈阳奥体万达广场

开业时间	2013 / 07 / 26
建设地点	辽宁 / 沈阳
占地面积	7.11 公顷
建筑面积	56.33 万平方米

OPENED ON	JULY 26 / 2013
LOCATION	SHENYANG / LIAONING PROVINCE
LAND AREA	7.11 HECTARES
FLOOR AREA	563,300 m²

OVERVIEW OF PLAZA
广场概述

沈阳奥体万达广场地处沈阳浑南新区浑河南岸，紧临沈阳奥体中心——2013年全运会的主场馆。整个用地为狭长的L形场地。规划上结合浑南新区的城市设计，沿浑河大街方向布置超高层城市豪宅，沿浑南中路一侧规划购物中心及写字楼。在两条路交会的地方设置了超五星级酒店——沈阳万达文华酒店。

Shenyang Olympic Wanda Plaza, located on the south bank of Hun River, Hunnan New District of Shenyang, is adjacent to Shenyang Olympic Sports Center Stadium which is also the main stadium of National Games 2013. The whole land is a narrow L shape. In accordance with the city planning of Hunnan New District, high-rise city mansion is arranged along the direction of Hunhe Street, and shopping centre and office building are arranged along Hunnan Middle Road. At the cross of the two roads Cocates the super 5-star hotel - Wanda Vista Shenyang.

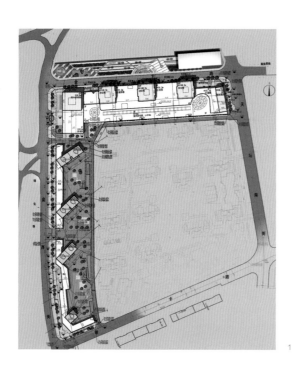

1　广场总平面图

2　广场外立面

3

FACADE OF PLAZA
广场外装

沈阳奥体万达广场以大气雄浑的整体气势, 搭建了一
个精彩纷呈的城市舞台, 在这个舞台上有活力四射的
商业综合体, 有尊贵典雅的超五星级酒店, 有高耸挺
拔的写字楼, 有气贯长虹的商业综合体, 还有温馨平
和的商务酒店。形态各异的建筑单体, 以符合各自功
能特点的外部形象独立存在, 同时又共同遵守着一
个合理而强大的内在逻辑, 从而形成一个和谐统一
的整体。

With the grand and vigorous atmosphere, Shenyang
Olympic Wanda Plaza builds a spectacular urban
theatre. On the stage, there is a vibrant commercial
complex, a noble and elegant super 5-star hotel,
a towering office building, a supreme and heroic
commercial complex as well as a cozy and warm
business hotel. A variety expression of individual
buildings represent their functional characteristics and
also comply under a rational logic, hence formed a
harmonious and unified entity.

3 广场外立面
4 广场外立面局部
5 广场立面图

4

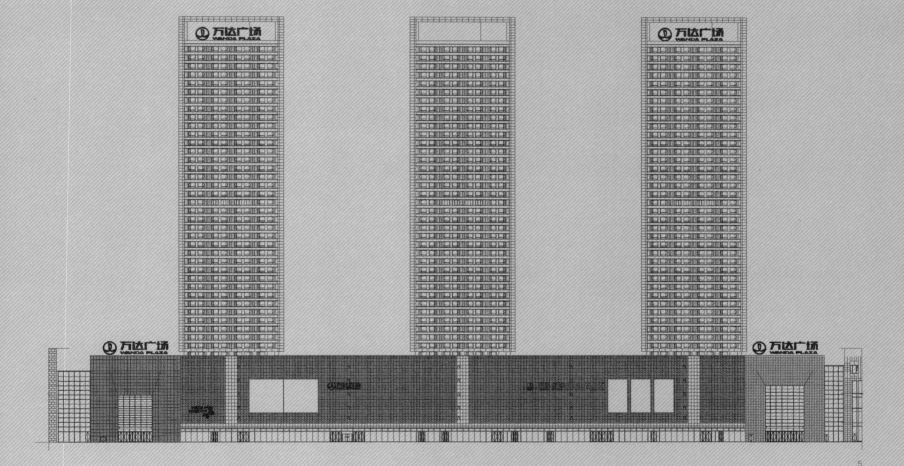

5

入口6层通高的大门采用双层表皮的手法，以丰富的空间层次凸显商业入口的热闹氛围，外层为点抓玻璃，内侧为1200毫米×1200毫米见方的中国红回纹凹凸透光板，双层的材质肌理彰显精致奢华的品质，也为项目增加了中国传统的商业韵味。

The 6-floor high entrance gate adopts double skin technology, with its rich space layers highlights the lively atmosphere of the entrance. The outer layer is point-supported glass system, and the inner layer is 1200mm×1200mm Chinese red concavo convex transparent plates. Double-layer materials and textures show the delicate and luxury quality, and increase the traditional commercial flavor of the project.

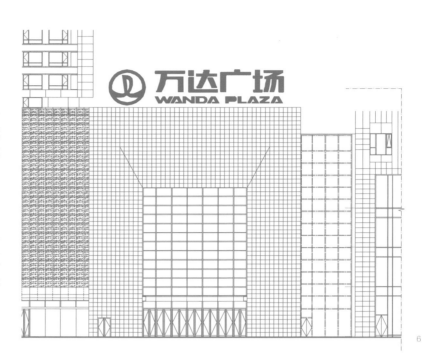

6

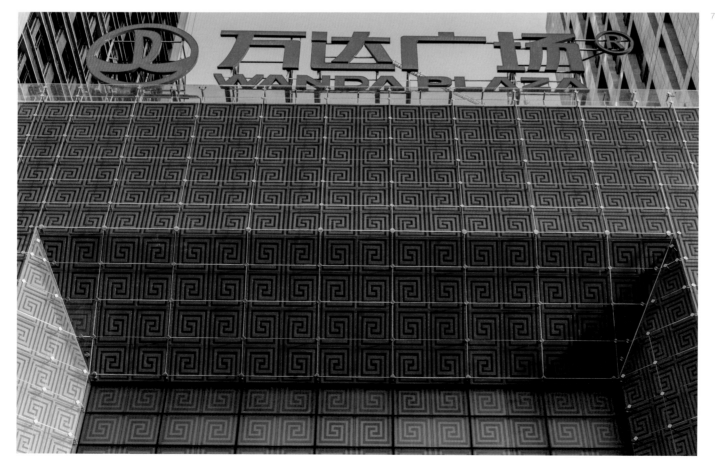

7

6 广场门头立面图
7 广场门头特写
8 广场一号入口

商业裙房采用层层出挑并向外倾斜的设计，更好地展现了外立面折型穿孔板丰富变化的肌理。同时，顶部采用弧形收头、雨棚处采用弧形甩尾，上下遥相呼应，构成了完整流畅、气势恢宏的商业裙房形象，为整个沈阳奥体万达广场项目奠定了端庄大气的格调。

The commercial podium is composed of cantelivered and layers tilting outward, which enhance the rich dynamic texture of the folding perforated plates applied on the facade. Meanwhile, the top portion used arch-shaped parapet and the curved canopy edge, echoes each other at a distance and completely compose a streamline and grand impression of the commercial podium. The whole project of Shenyang Olympic Wanda Plaza is therefore laid a dignified style.

INTERIOR OF PLAZA
广场内装

圆形成为贯穿整个购物中心的设计符号。无论是地面拼花,还是天花吊顶都依据圆形发生和展开;并通过圆形的汇聚与分散,来从视觉上标示出不同空间的层次地位,体现明晰的交通节点。在采光顶里,"圆"化身于一个个小孔,镶嵌在顶上的铝板形成的体块里,既起到了部分遮阳的功能,还产生了丰富的视觉光影。在大厅的设计中,它又化生为圆环,成为地面丰富变幻的图案,成为观光电梯立体动感的格栅。圆在这里旋转,圆在这里律动。长街天桥下,圆形冲孔图案,结合柔和温馨的灯光设计,构成了不同于以往任何项目的独特体量和别致美感。一个充满着青春气息,时尚品质的购物空间,就这样一点点,一幕幕在顾客的面前铺展开来。

Circular shape becomes the design symbol throughout the entire shopping centre. Both the parquet floor and the ceiling are developed on the basis of circular shape. The aggregation and dispersion of circular shapes visually indicates the hierarchy status of different spaces and clearly reflect traffic nodes. On the roof, "circular" becomes small holes embedded in the blocks made of aluminum plates, which plays partial shading function, meanwhile enriches visual effect of lights and shadows. In the hall, it becomes the rings as various patterns paved on the floor, or as the 3D dynamic grids in the sightseeing elevators. Circular shapes here are rotating and rhythmic. Under the over-bridge of the long street, circular perforate patterns with the combination of soft and warm lighting design, forms the unique model and the aesthetic feeling which no previous projects could compare to. A shopping space full of youthful and fashionable quality is therefore spreading in front of Customers.

10 椭圆中庭

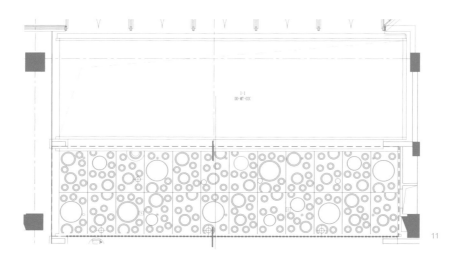

入口，作为室内与室外的衔接之处，有它特殊的地位。如引导或者说吸引室外的人们走进购物中心是我们特别关注的要点。浅咖啡色的背漆玻璃，与环状的光带、突起的不锈钢装饰灯具、墙面的透光石材一起赋予了入口动感时尚的效果。在这里，有一定反光的玻璃与透光石材，环状的平面图案与突起的圆形块面相互对比而又和谐共生。每当夜晚降临，入口丰富的光影，自然成为吸引室外人群的亮点。

The entrance, as the connection between the indoor and the outdoor, has its special status which guides or attracts people to walk into the shopping centre. Light brown back-painted glass, together with ring lights, protruded stainless steel decorative light fixture, and diaphanous stones on the wall, all give the effect of dynamic fashion to the entrance. At here, the glass with certain reflection and the diaphanous stones, plain circular patterns and protruded round blocks mutually contrast but coexist harmoniously. When night comes, the various lights at the entrance obviously become a highlight to attract the outdoor crowds.

11 主入口天花平面图
12 主入口门厅
13 主入口门厅特写

13

椭圆中庭采光顶的设计以太阳作为设计的原型，以白色软膜构成的放射状体块来象征太阳所散发的万丈光芒。恢宏的气势，在蓝天白云的背景下卓然彰显。侧裙板从整体方案的圆形符号出发，由大小不同的圆环，组合成丰富的图案，在白色基底的GRG的衬托下，动感跳跃，活力十足。这样的造型语言，再从侧面延续到顶面。在细部的处理上，结合灯光，形成圆环构筑的光带效果，让空间更加整体而生动。

The day lighting design of the elliptical atrium takes the sun as the prototype, with the radial shape made of white soft membranes symbolizing the radiant light emanating from the sun. The magnificent vigor is revealed under the background of blue sky and white clouds. Starting from the circular symbols of the theme, the side panels are combined with a variety of patterns formed by different sizes of rings. In contrast to the GRG white base it is so dynamic and energetic. This shape language is consistent from side to top. The detail treatment combines light effects to make the space more integrated and vivid.

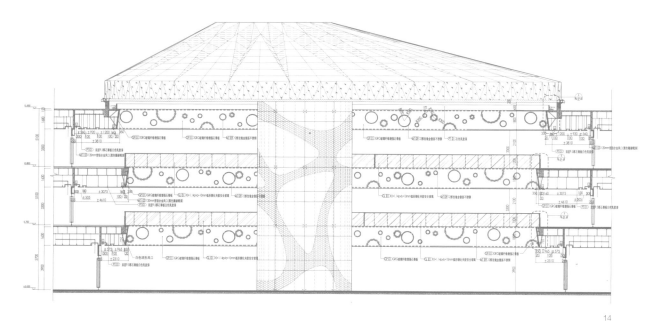

14

14 椭圆中庭剖面图
15 椭圆中庭

17

圆中庭的采光顶设计依然让整个空间仿佛处于一个
巨大太阳的照耀之下。玻璃与软膜，形成了刚与柔的
材质对比，更是带来了室内空间丰富的光线层次。圆
中庭侧裙造型，大小不一的圆点聚合，产生变化多端
又有模数可循的图案肌理。顶面发光软膜结合带有
圆点冲孔图案的铝板造型，构筑一条如星光璀璨的
光带，恍若银河。在有圆形图案格栅的观光梯内，仿
佛被圆形的气泡包围，则另有一番别样的体验。

In the elliptical atrium, the day lighting design of
the circular atrium makes the impression that the
entire space is under the huge sunshine. Glass and
membranes form the contrast of rigid and soft, and
provide the rich light layers to the indoor space. The
side panels gather different sizes of dots to create
varied but modular pattern textures. The top gleam
membranes combining with aluminum plate shape
which contains dot preforate patterns build up a starry
light stripe as the Galaxy. Inside sightseeing elevator
the circular-pattern grids feels like surrounded by
bubbles, which brings a unique experience.

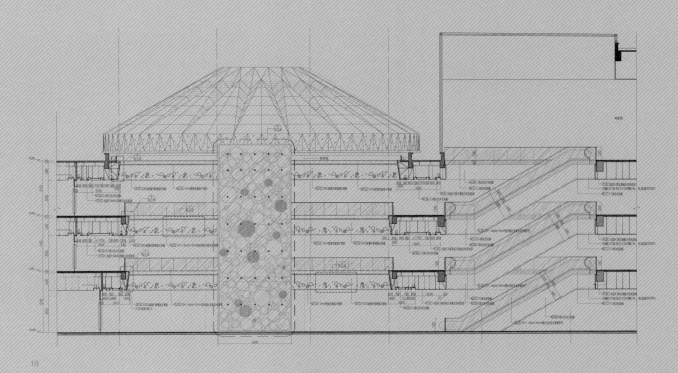

18

16 圆中庭采光顶
17 圆中庭
18 圆中庭剖面图

长街的设计采用将原来的洞口进行重新分组组合，再将天桥形状调整为具有宽窄变化的形体，使上层与下层之间构筑一种交错关系。这样的处理让长街的空间形态变得丰富。天桥弧度不同的外轮廓，使得天桥的两端宽窄不同。底部金属体块的设计，与发光圆点的图案，使得被加宽的天桥非但没有变得沉重，反而给人一种轻盈的感觉。

The design of the long street reassembly the original entrances and adjusts the overbridge to the form with width variation and build the alternating relation between the upper and lower. Such treatment enriches the spatial forms on the street. The outer contour with different radians of the overbridged makes its two ends have different widths. The metal block at the underside of the bridge, together with the luminous dot patterns, show a sense of lightness instead of making the widened bridge feels heavier.

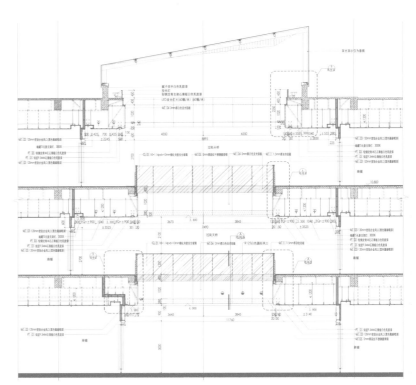

19

20

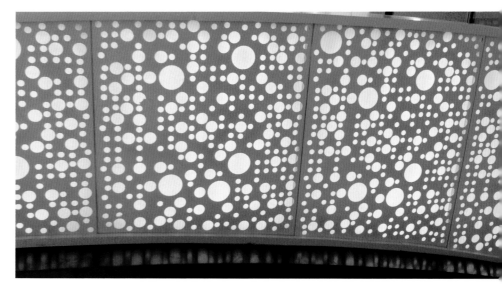

21

19 室内步行街剖面图
20 连桥
21 连桥天花
22 室内步行街

ONE STORE, ONE STYLE
一店一色

WANDA CINEMA / BIG STAR KTV
万达影城／大歌星

23a

23b

23c

24

23 广场景观主雕塑效果图
24 广场景观主雕塑
25 景观小品

LANDSCAPE OF PLAZA
广场景观

沈阳奥体万达广场景观设计力求打造一个高品质、精致、富有文化气息的城市综合体景观环境，将景观环境与建筑整体考虑，强调布局的整体感和细部的精致，同时营造出时尚、绚丽的夜景效果。

雕塑设计以水之环、运之城为主题进行设计，分为莫比乌斯环和冠军大道两部分。莫比乌斯环以第十二届全运会为主题，以环状进行主题的展示和与建筑形体的呼应。环高宽都为9米，以莫比乌斯环进行形体的扭曲变化，寓意在有限的空间做无限的运动，终点又成了新的起点。环的基座下有桥穿过环和水面，并有红外感应记录人的参与互动。

The landscape design of Shenyang Olympic Wanda Plaza seeks to create a high-quality and exquisite urban complex environment full of cultural atmosphere. It takes into account of the landscape environment and the overall architecture, emphasizes the integrity of the layout as well as the elegancy of the details, and meanwhile creates the stylish and gorgeous nightscape effects.

With the design theme of "rings of water, city of Games", the sculptures are divided into Mobius Ring and Champion Avenue. Mobius Ring takes the 12th National Games as the theme and echoes the architecture in the form of ring shape. With both height and width of 9 metre, Mobius Ring is twisted to change the form, implicate the infinite movement in the limited space while the end becoming a new beginning. A bridge and water come through under the ring base, and infrared monitors are applied to record people's participation and interaction.

25a

25b

25d

25c

NIGHTSCAPE OF PLAZA
广场夜景

通过对沈阳的历史人文、全运竞技等城市特色的提炼，设计出宏伟、大气、璀璨、靓丽高品位的夜景艺术亮点，在此基础上确立了以蜡烛塔楼为城市标志性建筑夜景的设计理念。大商业的照明采用柔和、淡雅的内透LED发光点，将灯具巧妙地与建筑结构相结合，达到见光不见灯的效果，在兼顾大商业白天景观效果的同时，增添了综合体整体的靓丽夜景效果。沈阳奥体万达广场整体夜景照明设计以绿色、科技、生态为核心，它的照明不再是以"亮"为目的，它的出现充分展示了浑南新区的新城市特质，提升了市民的自豪感和安全感，成为浑南新区夜景照明的主要观赏点。

The design is refined from the perspective of the city, including the history and humanities of Shenyang as well as athletics of the National Games, for a grand, gorgeous and sophisticated nightscape highlight. On this basis, the concept of "candle tower" as the landmark of the city is established. The commercial lighting adopts gentle and elegant interior LED luminous spots, which subtly combine with the architectural structure to achieve the effect of visible lighting, invisible light fixture. Considering the landscape effect in day time, it also enhances the beautiful night effects of the complex. Around the core of green, technology and ecology, the lighting design of Shenyang Olympic Wanda Plaza is no longer for the purpose of brightness. It fully shows the new urban characteristics of Hunnan New District, improves the citizens' pride and sense of security, hence becomes one of the main nightscapes in Hunnan New District.

FACADE OF HOTEL
酒店外装

沈阳万达文华酒店建筑外立面设计上采用大气端庄、简洁明快的手法,强调建筑造型与内部功能的完美统一。重视与城市环境的互动关系,使建筑与环境充分协调又独具个性。在建筑装饰细部的实现上,从结构功能的不同、装饰材料质感的把握、结合整栋建筑的风格,充分考虑到创意与工程实际的结合;并且通过对幕墙系统和结构的细部设计,确保幕墙的性能满足规范和设计要求。

Using dignified and concise techniques, the facade design of Wanda Vista focuses on the perfect integration of the architectural shape and the internal functions. It pays attention to the interaction with the urban environment and makes the architecture and the environment fully coordinated but respectively with personalities at the same time. In the realization of the decorative details, the design takes into account of the combination of creativity and practice in to different structural functions, textures of the decorative materials, and the overall architectural style. The design details of the curtain wall system and the structure ensure that the performance of the curtain wall meet the specifications and design requirements.

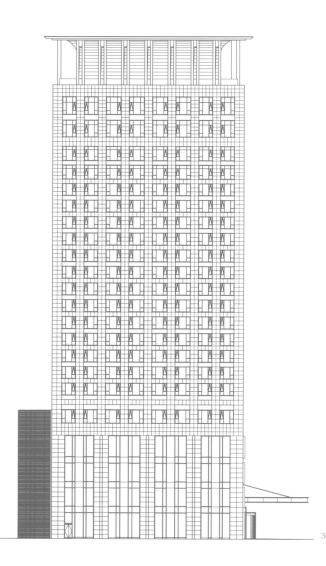

30

30 酒店北立面图
31 酒店主立面图

32

32 酒店入口
33 喷泉

33

LANDSCAPE OF HOTEL
酒店景观

酒店主入口以雨棚、入口水景、两侧景观墙和绿化形成独具酒店氛围的围合空间。铺地材料以黑色为主，间以黄色细条装饰形成精致细部并与建筑的立面分隔形成对应和延伸。入口水景以具有雕塑感的折面形体构成表面的肌理细节，两层的涌泉汩汩而下，构成独具特色的酒店标识。花坛矮墙中的景观灯柱采用与建筑呼应的形体风格设计，灯罩的细部、透光的云石以及精致的中国回纹都是与建筑细部有呼应的延伸，形成了酒店的独特标志景观元素。

Canopies, waterscape at the entrance and landscape walls on both sides form an enclosed space at the entrance with the unique hotel atmosphere. Paving material mainly in black, and decorated with yellow stripes, forms delicate details and extends correspondingly to the facade of the architecture. The sculptural folding sections form the texture details of the waterscape at the entrance. Two layers of fountain constitute a unique hotel logo. The landscape lights in the short walls of the flower bed adopt the design that reflects the style of architecture. The lampshade details, transparent marbles and delicate Chinese patterns all extends from the architecture details, which also formed the unique symbol of the landscape element.

34

35

34 喷泉

35 灯柱造型

36 景观水景

37 景观水景特写

38 绿化

36

37

38

NIGHTSCAPE OF HOTEL
酒店夜景

酒店主入口立面采用宽窄玻璃结合透光云石遮阳板的组合幕墙，玻璃及透光石材遮阳板在夜晚灯光下的错动组合形成了丰富的光影变化，衬托出文华酒店的精致和奢华。

The facade at the entrance adopts composite curtain wall glass with different widths and diaphanous marble sunshades. Combinations of the two materials under the lights at night form the varied lighting effects, emphasizing the refinement and luxury of Wanda Vista.

39 外立面夜景
40 入口夜景

XIAMEN JIMEI
WANDA PLAZA

厦门集美万达广场

开业时间 2013 / 06 / 08
建设地点 福建 / 厦门
占地面积 4.16 公顷
建筑面积 15.67 万平方米

OPENED ON JUNE 8 / 2013
LOCATION XIAMEN / FUJIAN PROVINCE
LAND AREA 4.16 HECTARES
FLOOR AREA 156,700 m²

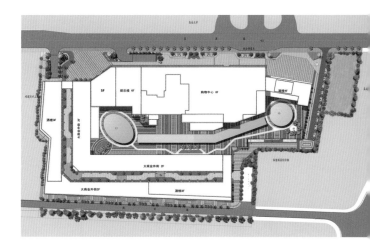

1

2

OVERVIEW OF PLAZA
广场概述

集美万达广场位于集美老城区同集路与集源路交会口，占地4.16万平方米，建筑面积15.67万平方米，建筑层数为6层，汇集万达百货、万达影城、大歌星KTV、大玩家电玩城、超市、大型酒楼等多个主力业态。该项目近距厦门大桥，紧邻集美大学以及集美长途汽车站，属于集美老城区内的黄金地段。

Jimei Wanda Plaza is located at the cross of Tongji Road and Jiyuan Road in old urban district of Jimei, with the land of 4.16 hectares and floor area of 156,700 m², 6 floors, including Wanda Department Store, Wanda Cinema, Big Star KTV, Super Player Center, supermarket, large restaurants and other programs. The project is near Xiamen Bridge and adjacent to Jimei University and Jimei Coach Station, which is at the prime locations in old urban area of Jimei District.

1 广场总平面图
2 广场全景图

3

FACADE OF PLAZA
广场外装

厦门集美万达广场的外立面整体设计思路是提取嘉庚风格的建筑元素,用现代的材料、工艺及表现手法进行创造组合,形成一种具有"新嘉庚风"特征的现代化商业立面,创造性地解决了商业建筑与地域特色相结合的问题。

The facade design concept of Xiamen Jimei Wanda Plaza is conceived from the architectural elements of Jiageng style. The creative combination of modern materials, techniques expression forms a modern commercial facade with the characteristics of "new Jiageng style", which innovatively solve the dilemma in which commercial architecture is combined with regional characteristics.

4

闽南建筑组群是以聚落的形式存在的，丰富的院落空间组织也是闽南建筑重要的特征之一。因此，基于这种建筑组群关系，在进行集美万达广场的造型设计时，可以很自然联想到错落有致的闽南传统聚落屋顶，从而对设计起到启发作用。利用大小不一的砖红色屋顶造型体将建筑形体进行覆盖，总计15块比例不一、形式略有不同的屋顶将原本略显单调的平屋顶进行了分割和组织，从而形成了丰富的屋顶造型，并使得建筑群具有了传统闽南民居聚落的韵味。

Minnan building group exists in the form of village settlement, and its important characteristic is the dynamic composition of the courtyard space. Based on the relationship of building group, the design massing of Jimei Wanda Plaza is inspired by the traditional well-proportioned settlement roofs. A total of 15 brick red roof shapes with different sizes and forms cover the architecture, in order break down and reorganize the originally boring and plain roof, and the rich roof style makes the building group have the charm of traditional neighbourhood.

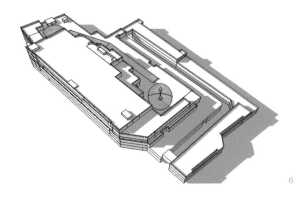

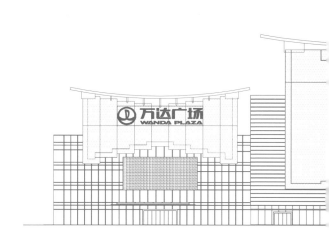

5 闽南建筑群
6 聚落结构模型
7 广场北立面
8 广场立面图

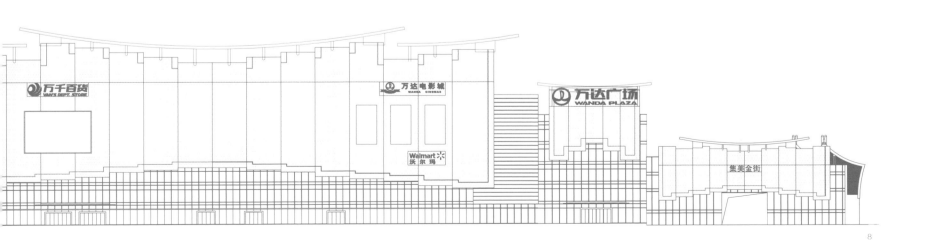

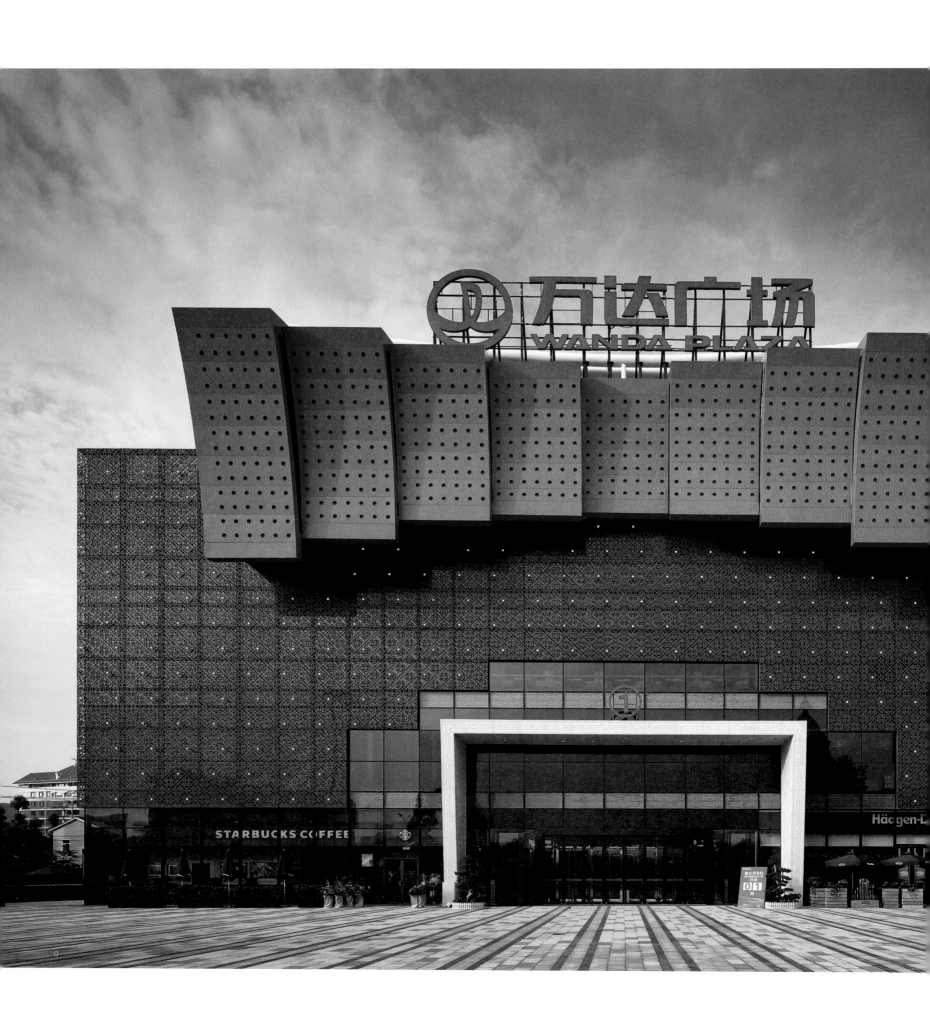

10

大商业入口是万达广场的第一入口。在这里，我们设
计了最具特点的闽南石浮雕元素的入口门套，凹凸有
致的万字纹图案搭配泉州白石材，使整个入口门套呈
现出高贵典雅的建筑形象。

The commercial entrance is the premium gateway of
Wanda plaza. The element of the most characteristic
Minnan stone reliefs on the door cover and the carved
cauighrphic "wan" patterns collocated with Quanzhou
white stones provide a noble and elegant architectural
image to the door cover.

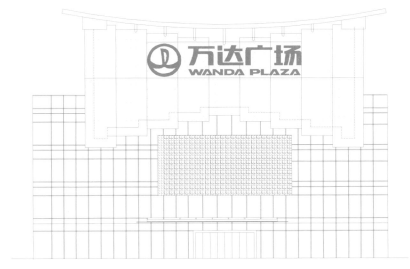

11

9　东入口
10　西入口
11　门头立面图

INTERIOR OF PLAZA
广场内装

厦门集美万达广场的外立面是极具地域文化的"嘉庚"风格,所以室内步行街的设计要点就是将建筑的形式"演变"到室内,并在室内空间的尺度上去体现建筑的美感。

同时考虑到在成本、工期等多方因素制约下的效果呈现,最终将设计主题的方向放在了通过对当地文化的提炼,形成一个可作为主题变化的图案上,并将该图案结合不同空间的尺度和位置特点进行平面及空间的演变,最终与纯粹的建筑空间进行叠加,以形成图案与空间的阴影关系,既本项目的室内设计主题——"案"之影!

The facade design of Xiamen Jimei Wanda Plaza contains rich local culture of "Jiageng" style. Therefore, the design key point of the indoor pedestrian street is to transform the architectural form to the interior and embody the aesthetic feeling in the scale of the interior space.

Meanwhile, considering the effect display under the limit of cost control, schedule and other factors, the design theme is eventually based on refining local cultures to form a changeable pattern as the theme. The plane and spatial evolutions of the pattern is combined with the characteristics of different space scales and locations finally overlay the pure architectural space to form the light and shadow relationship between the patterns and the space, which is the interior design theme of the project - shadow of patterns!

入口将地域文化与空间装饰紧密地结合在一起，尝试性地将当地材料（胭脂砖、影雕）与现代装饰材料（镜钢、灰镜）等有机地结合，使此空间不仅成为建筑空间上的室内外过渡，也成为视觉延续中一个过渡。

The entrance tightly combines the local culture to the space decoration. It attempts to organically combine local materials (rouge brick and shadow engraving) with modern decorative materials (glass steel and grey mirror), making the space not only the indoor and outdoor transition in terms of the architectural space, but also a transition in the visual continuation.

13

14

12　圆中庭采光顶
13　主入口门厅一
14　闽南文化砖墙
15　主入口门厅二

15

门厅天花将设计主题确定的图案通过丝网印的方式反映在了艺术玻璃上，并通过玫瑰金的大块面装饰板来与细腻的图案形成留白的视觉处理。同时地面的石材拼花巧妙地与天花造型形成了呼应及手法上的"对话"。

The foyer ceiling reflects the determined theme pattern on the art glass by means of screen print, and with the large rose gold decorative plates, the exquisite patterns form visual emptiness process. Meanwhile, the stone pavement cleverly reflects the shape on the ceiling by technical "dialogue".

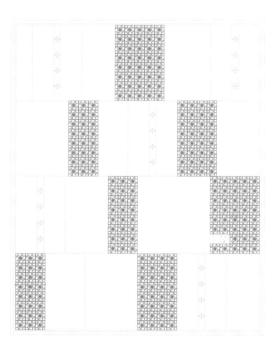

16

17

18

圆中庭将细腻、"多变"的视觉特性统一在了一个简约但不简单的空间中。通过既定的设计主题将可延伸的图案根据所在空间的尺度进行提炼，并在观光电梯的玻璃、天花的造型、挑空部分的侧裙以及地面拼花的图案上，通过设计手法附着的装饰材料的不同及特点呈现出丰富多变的视觉效果。

The circular atrium integrates the refined and verified visual characteristics into a minimalist but not simple space. Under the established design theme, the extended patterns are refined in terms of the scale of the space, then applied to the glass of the sightseeing elevators, ceiling shape, side skirts of the raised part and the floor pavement which show a variety of visual effects in accordance with different decorative materials and their features.

20

21

18　圆中庭采光顶
19　圆中庭
20　圆中庭侧裙板图案
21　圆中庭侧裙板

椭圆中庭通过既定的设计主题中图案的立体变化，来表达中国传统文化中的秩序之美。

By the spatial change of the established theme pattern, the elliptical atrium expresses the beauty in the order of the traditional Chinese style.

22

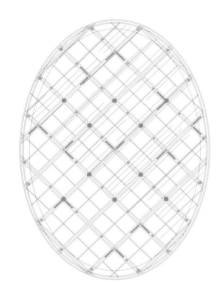

23

22 椭圆中庭采光顶平面图
23 椭圆中庭侧裙板
24 椭圆中庭

LANDSCAPE OF PLAZA
广场景观

厦门集美万达广场外环境延续建筑整体嘉庚风格，以嘉庚风为场地的大背景，同时挖掘集美当地的历史文脉，将其一个个历史点运用到节点空间来表达，最终形成了一条完整的故事线，使整个商业外广场有了完整的故事情节及文化内涵，也为整个项目营造了良好的室外环境。

主雕塑设计大胆运用了较为热烈的红色，红色在传统文化及商业氛围上找到一种平衡，同时母题选用大大小小的购物袋的设计，在购物袋上又融入了传统的镂空纹饰，使雕塑成为整个商业广场点睛之笔，使其成为传统与现代的完美融合。

The exterior environment of Xiamen Jimei Wanda Plaza is consistent with the overall "Jiageng style" of the architecture. With "Jiageng style" as the background, the landscape explores Jimei's historical context and applies the history points one by one to the key space nodes which finally complete a story line. The landscape gives the outdoor square a full story and the cultural connotation, and creates an excellent outdoor environment for the entire project.

The design of the main sculpture boldly uses the relatively rich warm red which balances the traditional culture and the commercial atmosphere. Meanwhile, the design in the form of big and small shopping bags, in which blends traditional hollow decorative patterns, highlights the sculpture from the plaza and makes the perfect combination of tradition and modern.

25

26

25 "剪纸"雕塑
26 "惠安女"雕塑
27 广场主雕塑

NIGHTSCAPE OF PLAZA
广场夜景

夜景照明设计凸显了嘉庚建筑的大屋面，屋顶构件飞脊，并用基础泛光照明表现了建筑体量。借助竖向光带表现立面幕墙的大体块分割，具有建筑的整体感又不失细部表现。设计方案真正做到灯光如水，如水随形，灯光依附建筑而生，充分表达建筑之美。

Nightscape design highlights the large roof and roof ridges of the Jiageng building, and basic floodlighting is adopted to present the architectural volume. Vertical light stripes generally segment the curtain wall of the facade, which contains integrality and also remains detail expression. The lighting program really achieved that the light is like water flow follows the form of the architecture, and fully expresses the beauty of the architecture.

29

30

29　外立面夜景
30　主入口夜景

OUTDOOR PEDESTRIAN STREET
室外步行街

室外步行街延续了立面的"嘉庚风格"主题，用艺术的手法、文化的精神、现代的表现力，通过建筑形体，景观小品的呼应化设计，渲染出浓郁的闽南文化气息。

Following the "Jiageng style" theme on the facade, the outdoor pedestrian street renders the rich Minnan culture in the form of architectural form and landscape treatments with artistic manners, cultural spirit and modern expression.

31

32

31 室外步行街外立面
32 室外步行街入口
33 室外步行街入口立面图

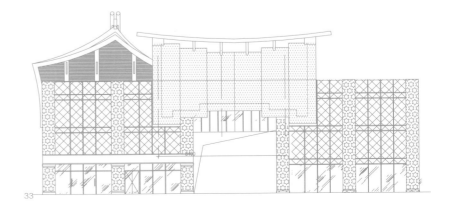

WUXI HUISHAN WANDA PLAZA

无锡惠山万达广场

开业时间	2013 / 06 / 21
建设地点	江苏 / 无锡
占地面积	8 公顷
建筑面积	36.06 万平方米

OPENED ON	JUNE 21 / 2013
LOCATION	WUXI / JIANGSU PROVINCE
LAND AREA	8.0 HECTARES
FLOOR AREA	360,600 m²

OVERVIEW OF PLAZA
广场概述

无锡惠山万达广场项目位于无锡市惠山经济开发区中心，占地面积8.0公顷，规划总建筑面积36.06万平方米。项目由购物中心、室外步行街、高级公寓及万达公馆等组成，涵盖购物、餐饮、休闲娱乐等功能。

Huishan Wanda Plaza is located at the center of Huishan Economic Development Area in Wuxi, occupyes land area of 8.0 hectares and total floor area of 360,600 m². The whole project consists of a shopping center, exterior commercial pedestrian street, superior apartment and Wanda Mansion, provides living, shopping, F&B, recreation, entertainment, and other programs.

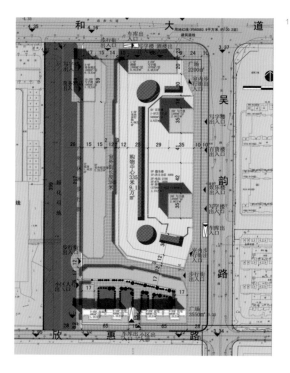

1

1　广场总平面图
2　广场外立面

FACADE OF PLAZA
广场外装

在规划设计上独创大商业多彩雕塑质感外立面，设计简洁却不简单；在广场主入口引入LED变换光源，最大限度地发挥空间尺度的美感，呈现前所未有的视觉冲击与色彩张力。设计力求营造出时尚、前卫、动感、趣味的购物空间。

利用富有体量感的三角体块形成立体雕塑群，倾斜动态的表面极具视觉张力。立面颜色选用代表丝绸感觉的紫红色做底色，隐喻无锡的繁花似锦。表皮配上白色线条，为夜景照明提供了安装条件，也丰富了底色。立面的彩色洞口，形成疏密有致的变化，内置彩色灯具，星星点点，犹如璀璨的星空，变幻闪耀。

Innovated on multi-color sculptural facade of the shopping centre, the design is simplified but not simple. The dynamic LED lighting source at the main entrance extents the aesthetic feeling of spatial grandness, and expressed an unprecedent visual impact and tension of color. The design aims to create a shopping space full of the sense of fashion, avent guard, dynamic and attraction.

Triangle blocks are utilized to form sculpture clusters with a rich sense of volume, and their inclined and dynamic surface creates an extremely strong visual impact. With its fuchsia undertone to implicate the sense of silk, the facade implies the flourishing scene of prosper Wuxi City. The surface of the facade is patterned with white strokes, making room for lighting fixtures, and enrich the undertone. The colored openings on the facade form well-proportioned variations, and by inserting lighting source inside, render the whole facade like a variable sky with twinkling stars.

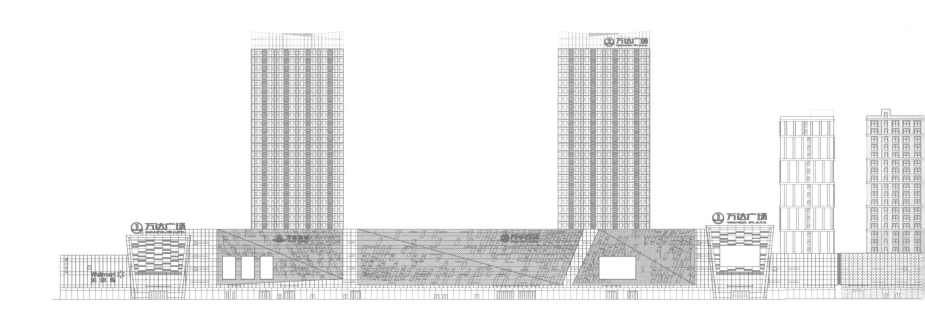

3 广场外立面
4 广场立面图

3

4

5

6

5 广场四号入口
6 广场主门头立面图
7 广场三号入口
8 广场二号入口

大商业主门头采用立体的玻璃盒子做法，折射环境色，玻璃表面有装饰感很强的彩釉图案，整体体块向外倾斜，富有张力，也强调了入口的可识别性，边框采用了双层玻璃的叠加效果，图案影像层次感分明，力图塑造商业高档精品店的炫酷效果。次门头是室外街通向中庭的重要入口，门头采用类似1、2号门头的做法，配以不同彩釉图案的玻璃，协调统一，与大商业立面相互呼应。

A method of making stereoscopic glass box has applied to the main gate of the shopping centre to refract the environment tone: the glass face is decorated with color ceramic glaze patterns, and the whole block is tilted outward with strong tension, makes the gate prominent and recognizable; the gate frame is overlapping with double-layer patterned, to demonstrate shining effect of supreme boutique shops. The secondary gate is an important entrance leading to the atrium, adopting similar methods for Gates 1 and 2 with multi-color ceramic glaze patterns to echo with the facade of the shopping centre in a harmony.

7

8

INTERIOR OF PLAZA
广场内装

通过波纹和菱形等设计元素的表现,体现其渔米之乡的独特文化底蕴,使作品更接地气。内装设计借助形体之间结构、材质、图案、灯光等细节的穿插变化,令空间极具细节品质,赋予了广场整体高品位的格调和惠山独有的性格。从大的空间形体刻画,及材质、肌理、色彩、灯光、节奏关系上处理变化,追求空间上的整体统一,细节上的刻画丰富,通体流畅的空间中结合结构的组合关系,使得室内的基础关系得以确定,并结合融入当地文化特色以及城市文化符号,让整体空间氛围和谐统一。

Designed with corrugated and rhombic elements, the exterior of the plaza presents the distinct cultural connotation of the city as a home of fish and rice, making the project more realistic. The interior design, in virtue of variations of structures, materials, patterns and lights in between the shapes, creates the space with refined details and endows the plaza a high level of taste and uniqueness. Through treatment of interrelationship of large spatial form, materials, patterns, colors, lights and rhythm, the project pursues for integrity of space and richness of details; also through interrelationship of combined structures, it makes the exterior more steady and adapted to the local cultural characteristics and cultural symbol of the city, enabling the overall space more harmonious.

入口处的设计考虑了形式与功能的结合，在纵向延伸的空间里设计了一组由灯光和镜面金属材质组合而成的天幕，钻石造型的元素从立面转折而入顶面，自成一体，尤为震撼。内藏光的灯箱以及有细节肌理纹样的浅香槟镜钢，组成虚实结合，闪烁有致的空间感受，恰似主入口的设计概念源于粼粼湖水泛起的波光，行走其间犹如踏入九天银河之中。

Consideration is also taken for the entrance design, in terms of integration of form and function: in the vertically extended space, a group of backdrops composed lights and metallic mirrors are installed, and diamond-shaped elements are extended from the facade to the top to make the whole face a strong visual shock. The light box and patterned stainless steel of light-champagne color make the space of both real and virtual, similar to the design concept of the main entrance which is inspired by the glittering water, makes one feel as to step into the Milky Way.

9 椭圆中庭
10 主入口天花

11

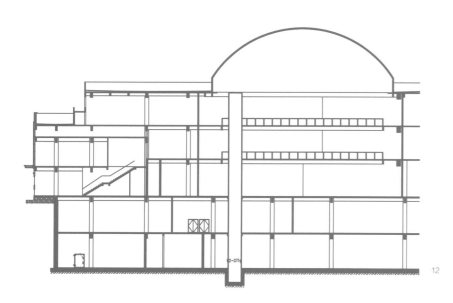

12

11 圆中庭
12 圆中庭剖面图
13 圆中庭侧裙板

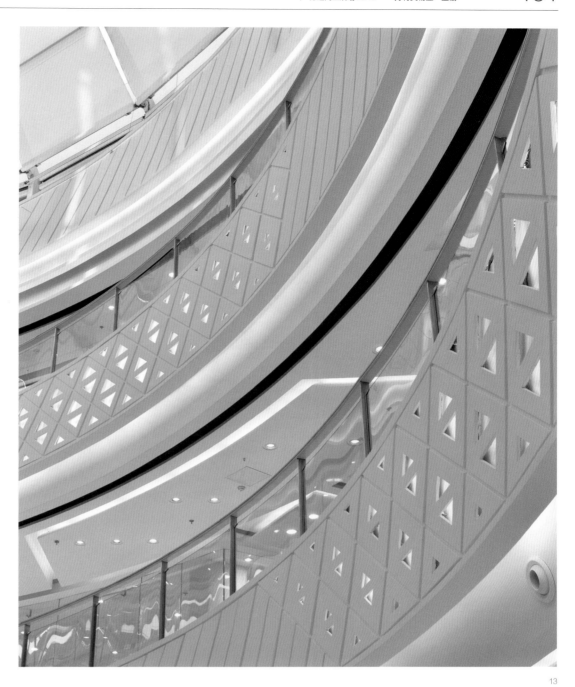

13

圆中庭也沿用虚实对比的设计理念，但手法却不同。此处采用的是一个凹孔藏光的造型，让底纹与造型互相融合，同时让光成为一种肌理元素体块，在空间中穿插，形成交错有序的视觉效果。圆中庭侧裙板均为统一的材料，但是通过造型与灯光，让其呈现出两种色彩、质感的感受，在阳光和灯光下呈现出一阴一阳的虚实变换，同时注重细节的刻画，以斜向线条为主调的设计元素有宽窄、长短、大小的变化，并同样注重模数化的设计根本，使得造型在施工的过程中准确、可控。

The round atrium also follows the design concept of comparison between reality and virtuality, but in a different way. What is applied here is a concave shape of light source, integrating the shading and shape and making the light a dynamic element to go through the space and form staggered and orderly visual effect. The side skirts of the atrium are made of uniform materials, while by means of shape and light, the side skirts present two different colors and quality, as if an exchange of reality and virtuality of Yin and Yang in the sunshine and light. The design also focuses on the details by making changes of width, length and size through the inclined stroke, and on the fundamental of modular conformation to enable the shape precise and controllable in the process of construction.

15

14 室内步行街仰视
15 室内步行街
16 室内步行街侧邦

16

直街的设计在空间中采用模数化的设计手法，通过设定可重复拼接延续的标准模块，融合了顶面造型、侧裙结构、地面拼花及机电设备。这几大装饰要点，按照一定的距离长度设置，并在空间中有序排列，使得直街空间达到井然有序，有纵深感的视觉效果。另外在直街空间中，让连桥以及扶梯成为空间的视觉焦点，通过自成一体的独立结构、浅香槟色系的金属材质、冲孔图案的图纹肌理和内藏灯光的设计效果，营造对比鲜明的空间视觉冲击效果。

By means of modular design in the space and setting of standard modules which are repetitively and continuously connected, the straight gallery integrates top shapes, side skirt structures, floor patterns and E&M equipment. All these decorative elements are arranged at a proper interval and order, making the gallery well-ordered and proportioned with an appropriate visual effect of depth. Besides, the overhead bridge and escalators become visual focuses in the space of the gallery, creating vivid comparison of visual impact through a series of independent structures, metal materials of light champagne color, perforated patterns and inserted light source.

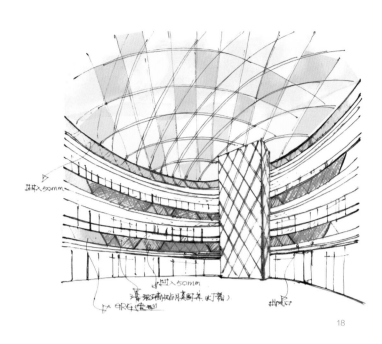

18

椭圆中庭采用了体块穿插的手法，让表现实体的GRG板以底纹的形式出现，让表现空间动感的镜面玻璃以灯箱的形式出现，虚实结合，形成交错有序的视觉效果。椭圆中庭侧裙板采用了香槟金色和白色两种色彩，并通过玻璃和GRG来分别呈现。模数块的造型尺度，均经历了反复的琢磨尝试，最终呈现出理想的视觉效果。圆中庭观光电梯采用了象征成熟稳重的藏蓝色色块，通体稳重大方，其细节亮点在于图案的处理方法，图案通过贴膜来实现，通过在玻璃的正面、背面错动拼贴带有反射效果的银膜，在人近尺度观察时会有多重反射的立体效果，远看又似晶莹剔透的宝石，带来美轮美奂的艺术效果。

The oval atrium applies alternative blocks to demonstrate the GRG boards in form of shading, show the dynamic mirror like finish in form of light box. In such combination, alternative and staggered visual effect is well presented. The side skirts of the oval atrium uses champagne and white colors through mirror like glass and GRG. The shape size of the module has gone through repetitive discussion and tests to present the final ideal visual effect. The panorama lift in the atrium applies dark blue block to show its mature and steady features. The details are highlighted by the patterns which are made of silver film staggered on the front and back sides of the glass which are reflective in close observation and look likes diamond when people stand afar, creating amazing artistic effect.

17　椭圆中庭
18　椭圆中概念草图

19

LANDSCAPE OF PLAZA
广场景观

无锡惠山万达广场像一个巨大的磁场，把人们聚集在了一起。而每座城市，都有属于它的时尚中心，每条街道，都有属于它的心情故事。如果说香榭丽舍大街和凯旋门广场诠释了法国的浪漫，那么惠山万达则集中了万达的智慧和惠山的奢华。"城市之心——代言都市生活"是本次景观设计之中心概念。根据万达商业综合体的特性，梳理功能分区明确景观定位。景观与建筑风格协调统一，典雅大方而又不失创意，氛围、尺度、细节舒适宜人。

Wuxi Huishan Wanda Plaza convenes people like a huge magnetic filed. Every city has its own fashion center, and every street has its own stories of mood. If we say the Champs Elysees and Arc de Triomphe interpret romance of France, then Huishan Wanda Plaza embodies the wisdom of Wanda and luxury of Huishan."Heart of City: Speaks for Metropolitan Life" is the central concept of the landscape design. In line with the characteristics of Wanda commercial complexes, the landscape design expressly defines the functional areas and makes it in harmonious coordination with the architectural style, presenting its elegance, grandness and innovativeness in dedicatedly designed dimension, details and atmosphere.

19 广场主雕塑
20 广场水景
21 广场绿化

20

水景设计同室外步行街入口相呼应，简洁的菱形边界同种植池相结合，特色植物向浅浅的镜面水默默倾诉。景观雕塑在水中悄然绽放，似花瓣也似花蕊，光影斑驳，唯美幽静。高大的建筑、蔚蓝的天空和娇艳的花朵在水波中荡漾，纯美自然。

The waterscape design corresponds to the exterior pedestrian street entrance: concise rhombic boundary combines with the planting bed, and feature plants are inclined to the shallow water face. Standing in water, the landscape sculpture looks like petals or buds, blooming in tranquil peace. Viewing the water, high buildings, blue sky and charming flowers are waving in rippling water, in the tremendously beautiful nature.

21

NIGHTSCAPE OF PLAZA
广场夜景

无锡惠山万达广场作为无锡市最重要的城市综合体，夜晚也将成为城市的夜间地标。照明设计从两个角度入手，第一要使夜景照明成为体现建筑、提升建筑的重要语言，充分理解建筑的设计理念和特点，使夜间照明及灯具在白天和夜晚都与建筑合为一体。第二通过灯光表现，使建筑成为表达无锡城市文化的载体，通过对城市文化的分析，对建筑在夜景上发挥"二次创造"的功能，强化建筑在夜晚的城市印象。

Being the most significant urban building complex of Wuxi, Huishan Wanda Plaza will be a landmark of the city together with its nightscape. The nightscape design sets out from two perspectives: first, making the nightscape lighting an important language of demonstrating and enhancing the building by fully understanding the architectural design concept and characteristics, and making the night lighting and fixtures integrate with the building both at night and in the daytime. Secondly, through light presentation, making the building a carrier of Wuxi's culture, and through analysis of the city's culture, exerting the "secondary creation" function of the nightscape for the building and consolidating the building's urban image at nights.

22

23

22 广场入口夜景
23 广场夜景

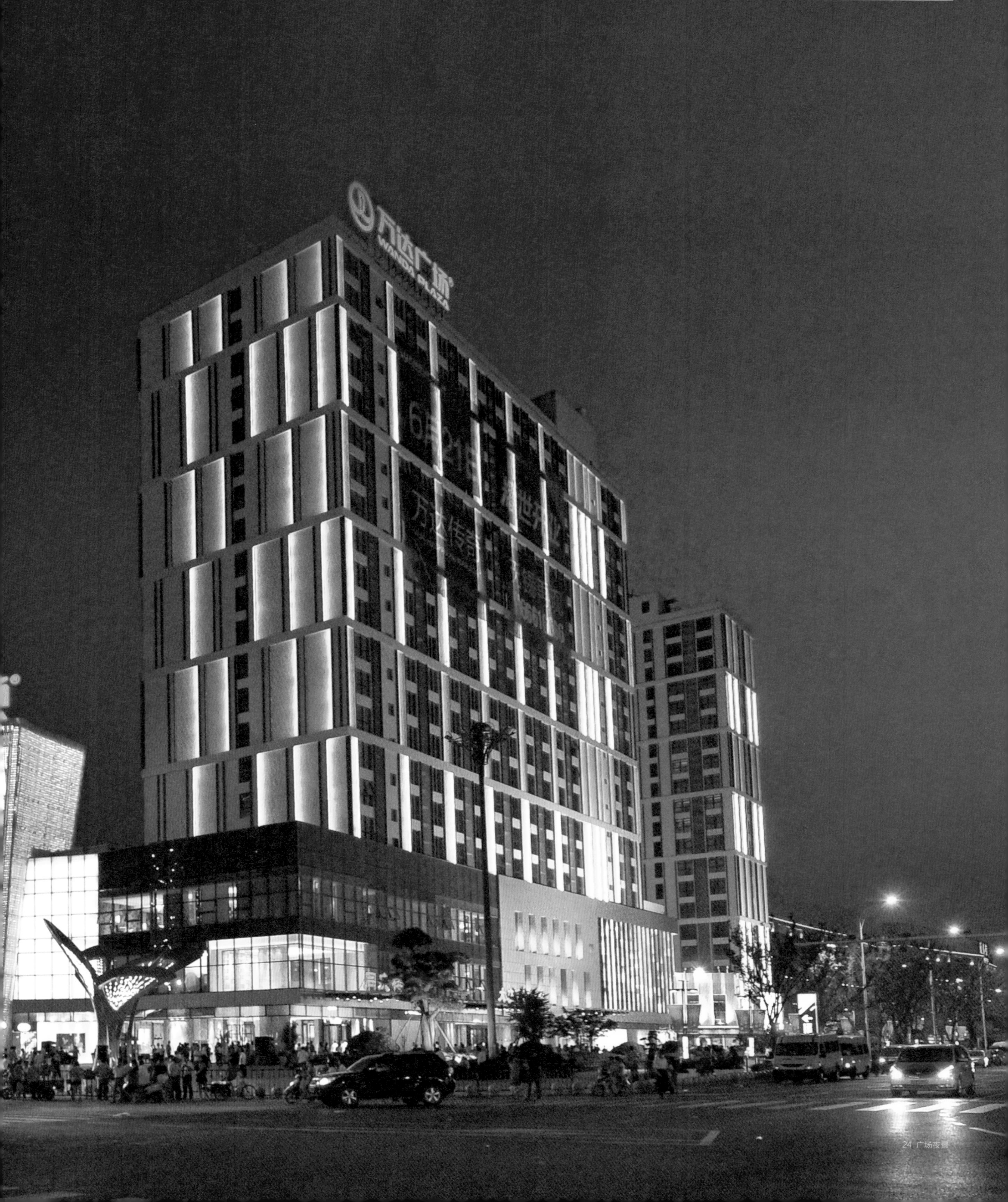

24 广场夜景

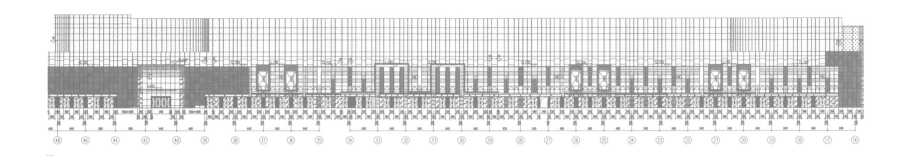

25

26

OUTDOOR PEDESTRIAN STREET
室外步行街

室外步行街主要营造一个浓厚的商业气氛空间。 主要有东南入口广场和西北入口广场这两个主要的入口，入口空间有"聚"的功能，除满足大量人群集散要求外，在此空间可进行多种多样的活动，既可成为商品展示的销售空间也让入口空间更具标志性与导向性，汇集大量的消费人群。

The commercial street is designed to create space with strong business and commercial atmosphere. The two main entrances are placed on the southeastern and northwestern sides of the plaza to "converge" people. In addition to such converging function, this space can also be venue for various types of activities, making the entrances a place to sell and display but also a symbolic and guiding sign to the inside of the plaza.

25　室外步行街立面图
26　室外步行街外立面
27　室外步行街雨棚
28　室外步行街景观小品

DONGGUAN CHANG'AN WANDA PLAZA
东莞长安万达广场

开业时间 2013 / 07 / 20
建设地点 广东 / 东莞
占地面积 8.85 公顷
建筑面积 37.25 万平方米

OPENED ON JULY 20 / 2013
LOCATION DONGGUAN / GUANGDONG PROVINCE
LAND AREA 8.85 HECTARES
FLOOR AREA 372,500 m²

OVERVIEW OF PLAZA
广场概述

东莞长安万达广场地处东莞市长安镇繁华核心区，基地环绕城市主干道，毗邻长安镇政府，紧临长安镇地标"长安门"，周边环境成熟，商业氛围浓厚，交通十分便捷，是万达集团在东莞地区投资建设的首个大型城市综合体，也是万达集团布局珠三角的又一力作。该项目由购物中心、商铺和精装住宅等组成，总建筑面积超过38万平方米，其中地上建筑面积约27万平方米，地下约11万平方米。

Chang'an Wanda Plaza is located at the bustling core area of Chang'an Town, Dongguan, close to the city artery, the township government of Chang'an, and the landmark "Chang'an Gate" of the town. The surrounding environment of the plaza is commercially developed with convenient traffic. Chang'an Wanda Plaza is the large urban building complex firstly invested and constructed in Dongguan by Wanda Group, and also another masterpiece of Wanda in the Pearl River Delta region. This project consists of a shopping center, boutique shops and luxury apartment, and its total floor area reaches over 380,000 m², of which that for above ground part is around 270,000 m², and that for basement is around 110,000 m².

1 广场鸟瞰图
2 广场总平面图

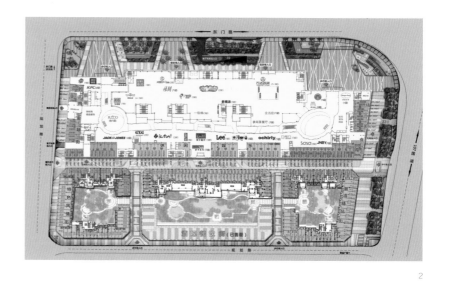

2

FACADE OF PLAZA
广场外装

东莞长安万达广场设计师深受著名建筑师圣地亚哥·卡拉特拉瓦（Santiago Calatrava）建筑思想的影响，试图模仿自然界蜜蜂翅膀的生动特点和天然钻石的造型。大商业裙房以"时尚之钻"作为主题，展现了时尚、现代的风格。"时尚芭比"为主题的外街，造型简洁大方，风格时尚活泼，一店一色，而酒店则采用最简洁明快的形式，完美地融入建筑群中。

外装设计注重大关系的把握，通过"大黄蜂"门头、"小黄蜂"段、"钢琴"段形成整体体量、材质、色彩、韵律的对比与和谐统一。大面积玻璃幕墙、多维幕墙、多层次透空门头等手法首次运用，完美实现"创新"思维。

Deeply influenced by the architectural concept of world-renowned architect Mr. Santiago Calatrava, the designer of Chang'an Wanda Plaza attempts to imitate the dynamic characteristics of wasp's wings and the shape of natural diamond. The podium of the large commercial area is a "diamond of fashion" to show its modern and fashionable style. The exterior street, themed with "Fashion Barbie", is concise in shape while lively in style, and each shop shows unique characteristic. The hotel part applies the simplest and concisest form to perfectly indulge in the building cluster.

The exterior design focuses on large relations in between the "Big Wasp" gate, "Small Wasp" block, and "Piano" blocks to form comparison and harmonious uniformness of mass, materials, colors and rhythms. The large glass facade, dimensional curtain wall and multi-layer perforated gates are applied for the first time to perfectly interpret "innovative" methodology.

3

3 广场外立面

商业裙房的外立面，着重展现东莞市的青春活力和经济实力，以"时尚之钻"为主题，设计的灵感来源于钻石形态，模仿其不同转折面的色彩和质地及其随着光影变化而变幻的特点。同时，这一造型的灵感来自"黄蜂"的翅膀特点，采用具有透明感和光泽度的彩釉玻璃作为材料，利用颜色深浅的变化和组合，动感十足，每个"黄蜂"翅膀造型都仿佛蜜蜂的两只翅膀在阳光下颤动。"黄蜂"翅膀造型同时有不同的变化，裙房展开面是"小黄蜂"的造型，而商业的两个最主要入口则采用"大黄蜂"的造型，它们在变化中又有很强的统一性。

The facade of the commercial podium aims to demonstrate the vigor of youth and economic strength of Dongguan. The design, themed with "Diamond of Fashion", sources its ideas from the diamond shape, and imitates colors and quality of different folded faces, and the variable features along with the light. In the meantime, this shape is also originated from the wasp's wings, and uses color ceramic glaze with gloss and transparency as material and changes and combination of different colors to make the shape look like wasp's wings flapping in the sun. The shape of wasp wings also enjoys variations, i.e. it appears a small wasp at the podium while a big wasp at the two main entrances of the commercial area, showing an integration of difference and uniformness.

4

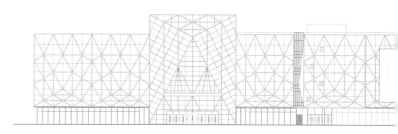

4　百货外立面
5　广场外立面
6　广场立面图

5

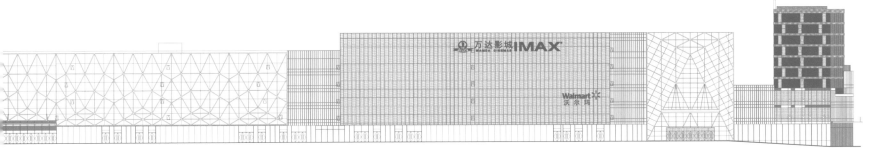

6

"大黄蜂"门头采用钻石形金属框架与玻璃幕墙，内衬三维钻石型穿孔铝板幕墙，建筑、夜景、内装多专业配合，营造浓郁商业氛围。

The "Big Wasp" gate employs diamond-shaped metal frame and glass curtain wall, lining with 3D diamond-shaped perforated aluminum plate curtain wall. In coordination with architectural, nightscape and interior design, it presents strong atmosphere of business.

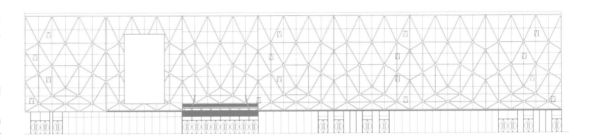

7

8

9

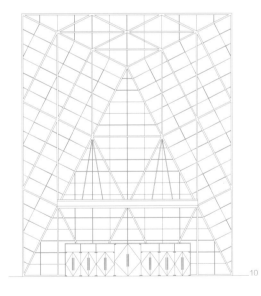

10

7　广场立面图
8　广场外立面
9　广场主门头立面图
10　广场主门头

11

12

"钢琴段"幕墙采用三维金属穿孔幕墙，自由舒展，与玻璃幕墙主体形成对比，成为形态色彩浪漫活跃的视觉焦点。幕墙与夜景有机结合，夜晚展现高精度动画效果。

The "Piano" block applies 3D metal perforated curtain wall in a free and flexible way, making comparison with the glass curtain wall and forming a visual focus which is colorful and dynamic. Through organic combination of curtain walls and nightscape, the project shows high definition animation effect at nights.

INTERIOR OF PLAZA
广场内装

东莞长安万达广场内装设计注重整体效果和建筑空间感的展现，运用钻石的元素为主题，设计手法简洁、流畅。此项目在直街侧裙板工艺上进行了创新突破，第一次使用曲面铝板的特殊工艺做法，使得直街段的效果非常整体连贯；同时，金属铝板的运用赋予了空间科技感和现代感，符合商业广场所在地——东莞的城市文化属性。直街的空间注重整体空间感的营造，二、三层连桥在内装设计时做了空间的错动和斜向改造，丰富了空间形态并使整个空间更加灵动与通透，产生"移步异景"的观感效果。椭圆中庭和圆中庭的采光顶设计进行了创新，在采光顶中心增加了别致的造型设计，结合采光顶的构造进行了内装效果的再次提升。

The interior design for Dongguan Chang'an Wanda Plaza emphasizes the presentation of integral effect and a sense of architectural space. Using diamond as its theme, the design is concise and fluent. The project is innovative in terms of the side plate process of the straight gallery, since for the first time, it applies the special process of curved surface aluminum plate to make the straight gallery quite continuous. Meanwhile, the application of metal aluminum plate endows the space a sense of high-tech and modern style, consistent with nature of the urban culture of Dongguan. The straight gallery space aims to create an integral sense of space: the interior design makes spatial staggered and inclined renovation for the overhead bridges on F2 and F3, which enriches the spatial shape and enables the whole space to be more flexible and transparent, producing a perception of "scenes vary along with proceedings". Innovation is also made for the skylight design of the oval and round atriums. Unconventional shapes are designed in the centers of the skylights, which secondarily enhance the interior effect in combination of the skylight conformation.

13 室内步行街

14

15

入口设计是内装设计的重中之重，为了给顾客营造最好的第一感觉，建筑形态以玻璃幕墙在入口处打造了挑空的过渡空间，内装设计考虑建筑外观的整体效果，运用建筑入口外观的"大黄蜂"折面与内装的斜线元素相融合，采用两种颜色的金属铝板形成立体化的造型，在彩色LED灯的照射下呈现变幻的色彩，形成大气、整体感的装饰图案。

The entrance design is a priority of the interior design. In order to create the best first image, at the entrance, an overhead transitional space is made in form of glass curtain wall, and by taking account of the overall effect of building appearance, the interior design integrates the "Big Wasp" folded face at the entrance with the interior inclined element, uses metal aluminum plates of two colors to form a 3D shape, and in virtue of colored LED, it presents changing colors and forming grand and integral decorative patterns.

14 主入口
15 主入口门厅
16 入口天花平面图
17 入口天花

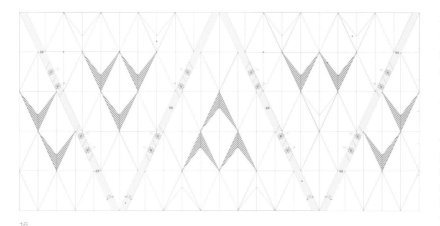

16

主入口门厅天花运用折面与斜线的混搭,采用简洁、立体的"钻石"艺术造型作为装饰元素,在一片洁白纯净的钻石体块中,偶尔点缀了几处闪烁有致的浅色香槟金与银色镜面不锈钢,组成虚实的凹凸体块,白色的立体造型在周边灯光渲染下呈现出丰富统一的艺术效果,提升了整个购物中心的品质感。

The lobby ceiling of the main entrance mixes the folded face and inclined line and uses the concise and dimensional shape of diamond as decorative elements. On the diamond block which appears white and clean, it is randomly dotted with light champagne color metal and silver mirror like finish, creating concave and convex blocks and making this dimensional white shape present rich and uniform artistic effect in the rendering of the peripheral lights. In this way the overall quality of the shopping center enhanced.

17

圆中庭注重的则是横向的发散变化，侧裙板立面设计在整体斜向条形造型的基础上用大小渐变的菱形块穿插其中，侧裙板光盒造型与外凸菱形块造型按模数各层错位布置，在纯白的中庭空间上，光盒造型相互穿插，形成发散并上下交错的视觉效果。整体风格简洁，设计手法轻松而流畅，注重整体效果的同时也不失对精致细节的刻画。

采光顶中心的铝板造型增加了商业空间的趣味性，与遮光帘组合形成了震撼的结构形态。整体色彩以白色为主，色彩统一，用灯光的冷暖变化营造高端时尚的商业氛围。

The round atrium focuses horizontal and extensive changes. The side skirts are designed with not only the inclined strokes but also the variably sized rhombic blocks interspersing the skirts. The light box shape of the side skirts and the convex rhombic blocks are alternatively arranged in module so that in the purely white atrium, the light boxes are staggered to form alternative visual effect. The overall style is simple and concise with flexible design methodology, and with both the integral effect and detailed being dedicatedly considered.

The aluminum plate in the center of the skylight adds interest to the commercial space, which forms a sensational structure shape together with the blinds. Basically in white, the variation of cold and warm light sources endows the commercial space with luxurious and stylish elements.

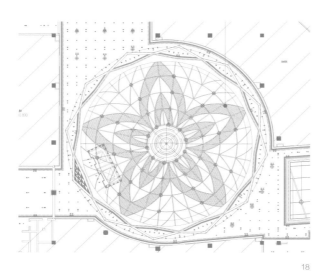

18

19

20

18　圆中庭采光顶平面图
19　圆中庭侧裙板
20　圆中庭

22

23

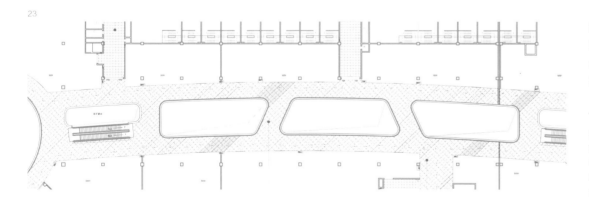

斜面塑造了步行街的性格，银灰色则是营造出了全场的格调。侧裙选用连贯的闪银色条形铝板作基底，偶尔间增加了几处柔和的发光玻璃或是镜钢面，营造出高端、时尚、现代的商业空间氛围。

The inclined face cultivates the features of the pedestrian street, and the silver gray is tone of the whole space. The side skirts apply continuous shining silver aluminum plates as base with random addition of gentle glass or mirror stainless steel, making the commercial space modern, luxurious and fashionable.

21 室内步行街仰视
22 室内步行街
23 室内步行街剖面图

椭圆中庭是一场"流星雨"式的演绎，洋洋洒洒的斜线与斜面，在偌大的中庭空间上显得极为精致与细腻。整个椭圆中庭采用宽窄有节奏的斜向的条形GRG板以及不锈钢板造型，使整个空间效果流畅而富有动感；地面材质在原定标方案基础上进一步丰富，用大块的颜色组合加强整体感。采光顶进行了适度创新，增加了别致的椭圆形造型，与周边遮光帘巧妙组合，打造出富有现代感的空间氛围；侧裙板造型采用宽窄不同的斜线勾勒出动感的空间形态。

The oval atrium interprets a scene of "meteor shower". The numerous inclined lines and faces appear delicate and refined in such a huge atrium space. The whole oval atrium applies GRG board with inclined strokes of different width and stainless steel plates, making the space dynamic and flexible; the flooring material is further enriched based on the original winning scheme, which uses huge color combination to strengthen the sense of integrity. Appropriate innovation is made for the skylight, which adds delicate oval shapes to skillfully match with the surrounding blinds and create a modern space. The side plate applies inclined strokes of different width to highlight the dynamic spatial form.

24

25

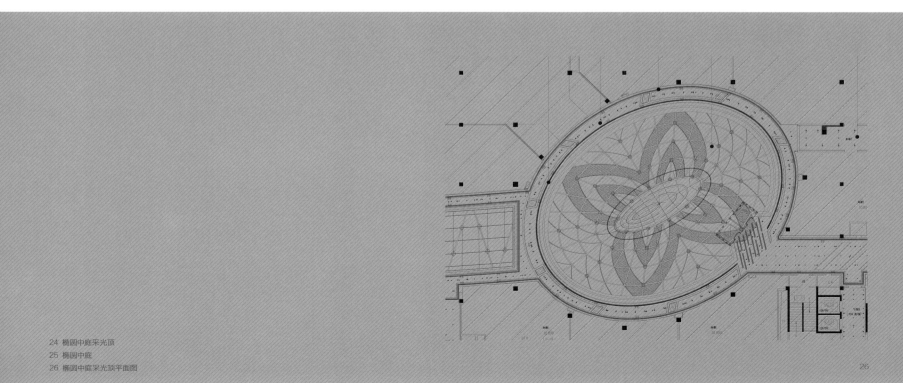

24 椭圆中庭采光顶
25 椭圆中庭
26 椭圆中庭采光顶平面图

26

LANDSCAPE OF PLAZA
广场景观

东莞长安万达广场景观在注重时尚与艺术的融合中，更加强调景观细节对人性的关怀。舒适的微地形空间、主题场景的营造以及台阶前提示地钉的设置都在真正关注人的需求，这是景观品质从形象品质到人文品质提升的体现。

In a combination of fashionable and artistic styles, the landscape design for Chang'an Wanda Plaza more emphasizes the humanistic concern from the landscape details. The comfortable micro topographic space, theme scene and the nails in front of the steps are showing the real concern about human's demands. This is improvement of landscape quality from image quality to humanistic quality.

27 广场花坛
28 广场绿化
29 广场台阶
30 水景

NIGHTSCAPE OF PLAZA
广场夜景

夜景照明设计充分考虑了项目的定位、建筑形态、地域特点。突出建筑特质，提升城市形象，体现当地文化，营造商业气氛。动画设计主题寓意希望，龙马精神，奔腾长安。

The nightscape lighting design fully takes account of the project orientation, building shape and regional characteristics. Such design also highlights the building features and improves the urban image to demonstrate the local culture and build strong commercial atmosphere. The theme of the animation design aims to place wishes on vigorous spirit of Chang'an.

31 广场立面夜景
32 广场门头夜景

31

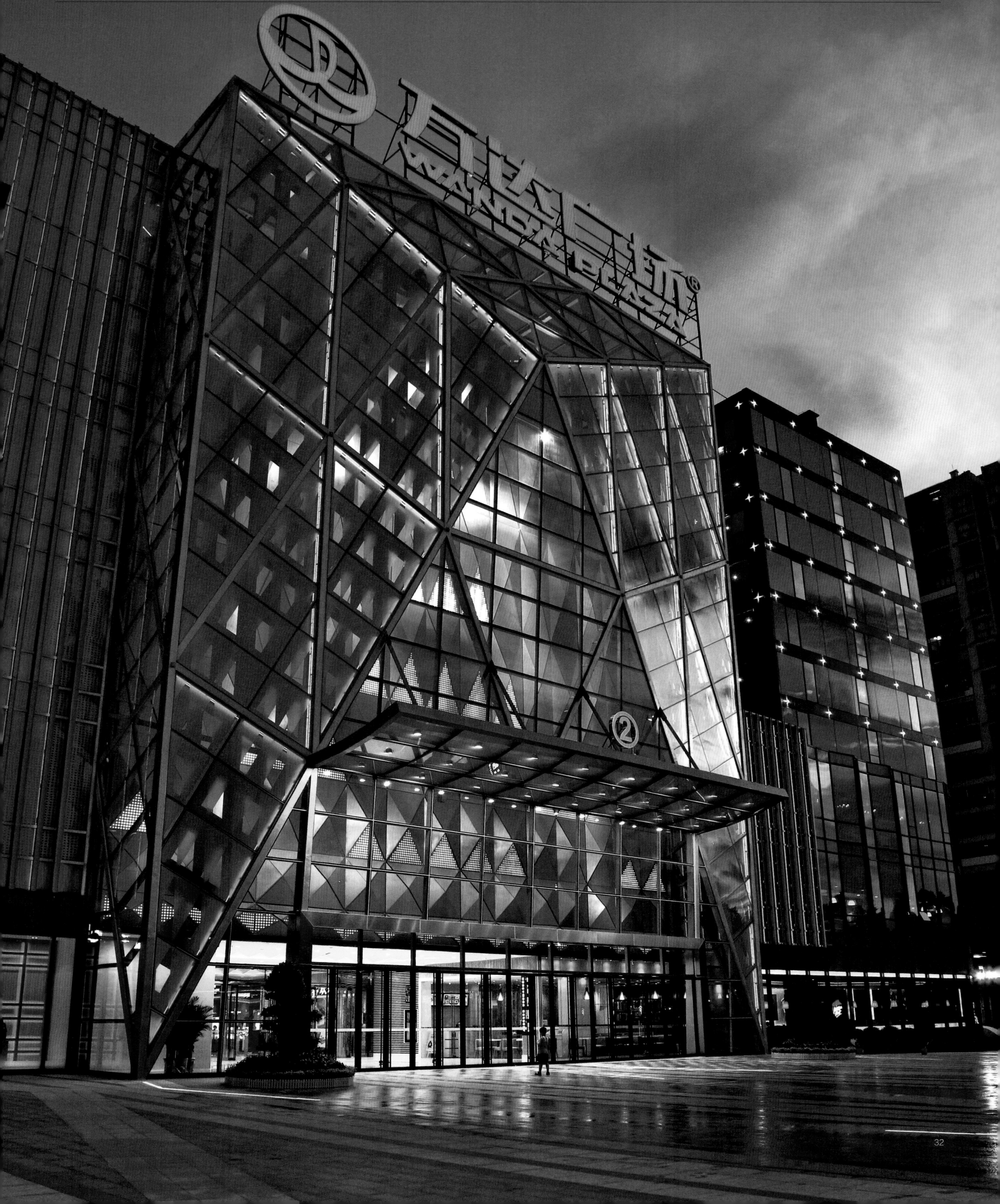
万达广场
WANDA PLAZA
2

33.广场夜景

OUTDOOR PEDESTRIAN STREET
室外步行街

室外步行街尺度宜人，单元形态多样，穿插组合；立面色彩、材质、空间变化丰富，结合夜景照明、景观美陈，营造充满活力的"一店一色"室外商业街。

The exterior commercial street is appropriate in dimension with diverse unit forms and alternative combinations. The facade colors, materials and space have different changes, and in coordination with the night lighting and landscape displays, it creates a vigorous commercial street with each shop showing unique characteristic.

34

35

34 室外步行街立面
35 室外步行街入口

36

37

38

36 室外步行街夜景
37 照明设计
38 橱窗展示
39 "芭比系列"雕塑
40 景观小品

39a

39b

40

39c

CHANGCHUN KUANCHENG WANDA PLAZA
长春宽城万达广场

开业时间   2013 / 08 / 16
建设地点   吉林 / 长春
占地面积   10.23 公顷
建筑面积   45.03 万平方米

OPENED ON   AUGUST 16 / 2013
LOCATION   CHANGCHUN / JILIN PROVINCE
LAND AREA   10.23 HECTARES
FLOOR AREA   450,300 m²

OVERVIEW OF PLAZA
广场概述

广场地处长春市宽城区，紧邻长春市第一主干道人民大街，毗邻长春市重要的交通枢纽长春火车站，交通便利，人气旺盛，是继长春重庆路万达广场、红旗街万达广场之后，万达集团在长春市贡献的又一力作。

长春宽城万达广场总用地10.23公顷，总建筑面积45.03万平方米。其中购物中心地上建筑面积9.36万平方米，地下7.07万平方米，建筑层数为5层，涵盖了万达百货、万达影城、大歌星KTV、大玩家电玩城、超市、大型酒楼和健身等多个主力业态。长春宽城万达广场的开业，也弥补了宽城区大型城市综合体的空白。

Kuancheng Wanda Plaza is located in Kuancheng District, Changchun, adjacent to Renmin Avenue, the first artery in Changchun City, and close to the important traffic hub of Changchun - Changchun Railway Station. On this location where is vigorous and convenient in traffic, Wanda Group contributes to Changchun another masterpiece following its previous two plazas in Changchun, namely Wanda Plazas at Chongqing Road and Hongqi Road.

Kuancheng Wanda Plaza occupies total land area of 10.23 hectares and total floor area of 450,300 m². Its shopping center has above ground floor area of 93,600 m² and underground floor area of 70,700 m². The building has five floors, accommodating Wanda Department Store, Wanda Cinema, Big star KTV, Super Player Center, supermarket, large restaurants and fitness center, etc. The opening of Kuancheng Wanda Plaza fills up void of large urban building complex in Kuancheng District.

2

1 广场总平面图
2 广场鸟瞰图

1

3

FACADE OF PLAZA
广场外装

长春宽城万达广场外立面的特点可以概括为：一轴长卷、两幅版画、三块浮雕、四幕彩绘。

一轴长卷使指长春宽城万达广场外立面由胶片的感觉衍生出的一个个折板单位组成，每一块折板都不垂直地面，通过折板之间的交接形成了连绵不断的一轴电影长卷。折板之间的每一个交点出挑首层橱窗的距离都不一样，最大达到4.5米，最小不到800mm，使得整个立面的点、线、面、体都有强烈的视觉冲击力。

两幅版画是指购物中心的两个步行街门头。作为整个购物中心的重要节点，将其处理成向城市展示的超大橱窗，最大门头宽70米，高30米。门头的处理手法为内衬一层彩色装饰背板，将电影元素提炼为抽象图案，在背板上用百页拼出，产生版画的效果。百页预留藏灯的条件，将灯具安装在百页中，夜晚形成一个巨大的LED屏幕。

3 广场外立面
4 广场立面图
5 主入口

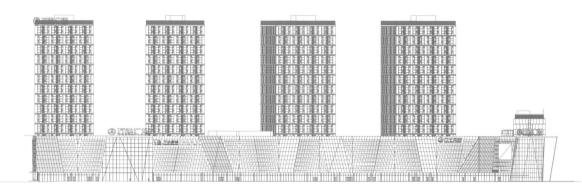

4

The facade of Kuancheng Wanda Plaza can be summarized as: a long scroll, two printmakings, three relieves, and four colored paintings. A long scroll refers to the numerous folded units deriving from the sense of films on the facade of the plaza. Each folded plate, rather than being vertical to the ground, forms a continuous long scroll of film by interfacing with another plate. Each intercrossing between two folded plates extrudes the GF show window at different height, the highest point reaches 4.5m, while the lowest point is less than 800m. In such arrangement, the whole facade presents tremendously strong visual impact in the combination of point, line, face and body.

The two printmakings refer to the two gates at the pedestrian streets. As major nodes of the whole shopping center, the two gates are designed as two super large shop windows displaying to the city. The largest gate is 70m wide and 30m high. The gates are lined with a layer of color base plate, on which film elements are refined as abstract patterns to be placed on the base plate with blinds to produce the effect of printmaking. Inside the blinds conditions for lights are reserved to form a huge LED screen at nights.

5

6

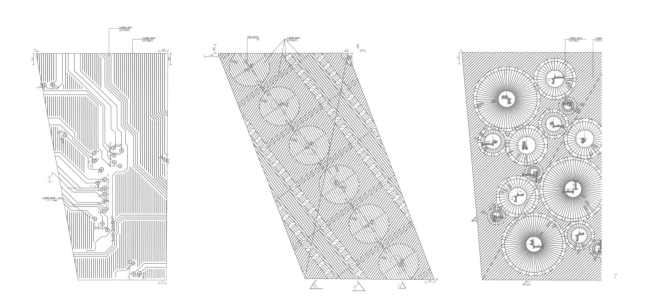

7

6 广场外立面
7 外立面图案平面图
8 浮雕特写

三块浮雕是指立面上展示出代表了长春当地文化特
色的三块浮雕，分别代表长春的电影文化、光电技术
和汽车文化。浮雕的构造节点分为两层，内侧为百页
密拼的图案，外侧为超白玻璃幕墙。浮雕面对城市主
干道，醒目、精致、秀美。它和城市形成了互动，将项
目融入城市当中。该项目为万达广场中外立面上有浮
雕处理手法的第一个项目。

The three relieves indicates the ones on the facade
which symbolize the local cultural characteristics of
Changchun: film culture, photoelectric technology,
and automobile culture of Changchun. The relieves
are divided into two layers, the inner layer are patterns
made of blinds, and the outer layer is super white
glass curtain wall. The relieves are facing the artery of
the city, appearing prominent, delicate and graceful.
Through this way, the project interacts with the city
and involves itself in the city's context. Kuancheng
Wanda Plaza is the first among Wanda plazas which
applies the treatment of relieves on the facade.

8

9

四幕彩绘是指四个25米高的广告幕墙，与建筑立面紧密融为一体，将购物中心的商业氛围提升到新的高度。四个广告位分处不同的位置，每个位置的设置都经过精心推敲。东南转折的楔形广告位，起着极强的引导作用，将东侧大广场的人流向南侧带形广场引导；百货入口处的广告位，将百货的主入口明确的强调出来；南立面上最西侧的广告位，临近城市主路凯旋路，路上的行人和车辆均能看到此超大广告；金街入口处的广告位，将金街入口体现得更加醒目，此处广告位给影城专用，展示内容为电影大片的宣传海报，将电影主题的设计理念更加深刻地植入每一个部分。广告幕墙采用超白玻璃幕墙，除了每层层间有拉结外，没有别的视线遮挡，为立面的商业感觉做出极大的贡献，并且因每个广告幕墙都为不规则形的空间面，广告幕墙本身也是一个让人群欣赏的设计元素。

The four colored paintings refer to the four 25m high advertisement curtain walls which are closely connected with the building facade and enhance the commercial atmosphere of the shopping center to a higher level. The four ad walls are positioning at different spots which are delicately designed and selected. The wedge-shaped ad wall on the southeast corner is a strong guide introducing the walk flow from the east large plaza to the south strip-shaped plaza; the ad wall at the department store entrance expressly emphasizes the main entrance; the ad wall on the far west on the south facade which is close to the city's main road Kaixuan Road, attracts attentions of pedestrians and cars to see such super large ad wall; the ad wall at the commercial entry makes this entry more prominent to people which is dedicated for the cinema to display large posters of films. This ad wall is deeply implanted with details of design concept of the film theme. The ad curtain walls apply super white glass with no visual barrier except for the joints between two layers, producing tremendous effect for the sense of commerce of the facade. Besides, as each ad wall has irregular spatial face, the ad wall itself becomes desirable design elements.

10a

10b

10c

9 百货外立面
10 外立面广告位

INTERIOR OF PLAZA
广场内装

长春宽城万达广场室内步行街内装设计着重于建筑整体的思考，深入研究，力求设计出与企业文化和地域文化相吻合的空间形象。在设计中以三角抽象的燕尾形为主要设计元素，寓意蒸蒸日上的商机，整体设计在结合形体、材质、肌理、色彩和灯光等变化的同时，又考虑了外观设计的圆形渐变设计元素，轻重结合，创造了一个造型简约、积极向上、内外统一的人文化商业环境。

The interior design for the gallery of Kuancheng Wanda Plaza focuses on overall consideration of the building in an attempt to integrate the corporate culture and the local culture. An abstract swallow-tail shape is designed to imply prosperous business opportunities. In the meantime of coordinating variations of shapes, materials, veins, colors and lights, it also takes into account of the gradual change of round shape on the appearance to create a concise, vigorous and uniform cultural and commercial environment.

11

11 圆中庭概念手稿
12 椭圆中庭
13 椭圆中庭采光顶
14 椭圆中庭剖面图

12

椭圆中庭观光电梯的立面采用上下贯通的明露金属线条，并用光带加以强化，同时加上玻璃面层反射与透射的强烈对比，更强化了观光电梯作为视觉中心的感觉；再配以地面深浅色块穿插形成的大面积燕尾形组合图案做呼应，形成强烈的积极向上的动感。而侧裙板的处理则以圆形渐变的手法，加以光影的变化，与建筑外立面遥相呼应，达到内外统一。

Elevation of the panoramic lifts in the oval atrium apply open metal strips connecting bottom to top. Strengthened with light and strong comparison of reflection and projection of glass face, the panoramic lifts become a visual center of the atrium. Besides, alternated by large size swallow-tail shaped patterns on the floor, it produces a strong and dynamic sense. The gradual change of round on the side skirts, together with light changes, echoes with the building facade in a uniform way.

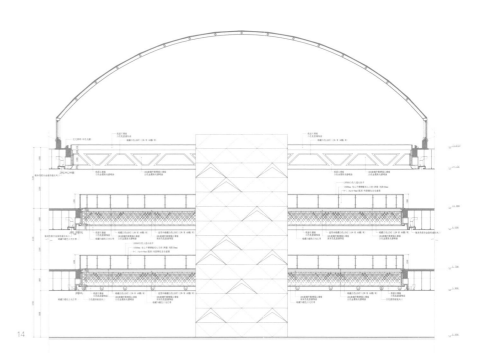

圆中庭的设计充分运用了建筑本身的圆形结构，地面以观光电梯为中心，以深浅组合的弧形石材拼条向外发散，形成美丽的涟漪图案，也强化了圆中庭的进深感。一、二层的天花侧裙板采用大面的折形斜面GRG造型，板面以三角形为基本元素，辅以凹、凸等手法，极大地丰富了圆中庭的空间视觉感受。而顶层女儿墙的侧裙板处理则由于视距较远，采取了弱化处理，以有节奏的凹槽环通，并局部加以渐变的组合灯盒作为点缀，手法简单，视觉感受却并不粗糙。观光电梯的立面采用大面的弧形玻璃，玻璃面层用反射与透射的处理手法形成向上的燕尾图案。

The round atrium design fully employs the circular structure of the building body, centers on the panoramic lifts on GF, and extends outwards with the arc stone bar in different depth to form beautiful ripple patterns and strengthen the sense of depth of the round atrium. The side skirts of F1 and F2 use large size folded GRG boards with inclined face; basically in triangle, the boards are arranged in convex and concave ways, which greatly enrich the visual effect of the round atrium. While the side skirts on the top parapet wall, due to far visual distance, apply weakening treatment with concave slot connection and locally gradual changes of light boxes to dot with the visual sense. The elevation of the panoramic lifts uses large size arc glass and forms flying swallow-tail shaped patterns by means of reflection and projection of the glass face.

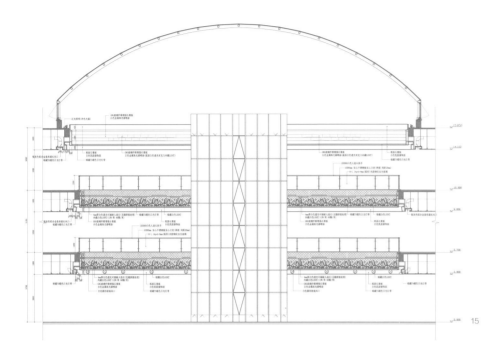

15

16

17a

入口门厅天花采用香槟金铝板的模数化的立体燕尾灯带造型，结合铺地中以直线条分割形成的深浅块面组合，形成强烈的导向性的视觉效果。

The ceiling of the entrance lobby applies modular 3D swallow-shaped lamps made of champaign gold aluminum plate. And in coordination with the blocks of different depth on the floor which are divided in straight lines, it forms strong visual effect for guiding directions.

17b

15 椭圆中庭剖面图
16 圆中庭
17 主入口天花

直街的自动扶梯与连桥在侧裙板与底板的设计上采用造型连通的形式，加强了自动扶梯与连桥的一体化感觉，再加上防护栏板的渐变图案设计，形成了直街强烈的空间艺术效果。而仅有的一个独立连桥则采用侧裙板、底板弧形一体的设计，饰面以砂面和镜面不锈钢相结合的纽带形式，纯手工打造，使独立连桥形成直街中另一个绚丽的视觉亮点，同时体现出长春城市仍具有的工业遗迹性。

The escalators and overhead bridge of the straight gallery employ connective and continuous form in their side and base plates, which strengthen integration of escalators and bridge. Besides, the gradual changes of guard boards are dedicatedly designed to produce strong artistic effect for the straight gallery. The separate overhead bridge applies an integration of side and base plates in an arc form, and an integration of frosted face and mirror stainless steel in a link form, totally manually, to make the overhead bridge another visual highlight of the straight gallery, and to demonstrate the industrial feature of Changchun.

18 室内步行街
19 连桥

19

ONE STORE, ONE STYLE
一店一色

20a

20b

20c

20d

LANDSCAPE OF PLAZA
广场景观

长春的电影业深刻影响了新中国一代人的成长，塑造了无数的英雄形象，培养了人们的爱国主义情愫，丰富了全国人民的精神文化生活。故景观设计除与建筑相互呼应、衬托以外，与当地电影文化相结合，采用"城市电影"这个主题。通过对这一主题的深入解读，利用现代景观的处理手法，打造出具有现代感、科技感、文化性的城市商业新环境。设计中融入长春电影文化特质，为人们精心打造现代城市新生活，体现新兴商业娱乐及情调生活。

The movie industry of Changchun has fundamentally influenced the generations of new China, which has created numerous heroes, cultivated patriotism and enriched spiritual and cultural life of the Chinese people. The landscape design for the project, in addition to echoing the building itself, the landscape design also combines with the local film culture and utilizes the theme of "city movie", through which deeply interprets the theme and uses modern landscaping method to build an urban new commercial environment with modern, high-tech and cultural characteristics. The design introduces the film culture of Changchun and dedicatedly builds a new life in the modern city the showcase emerging commercial entertainment and a life of new taste.

21 景观主雕塑
22 广场花坛
23 广场台阶
24 广场绿化

21

22

23

25a

25b

25c

25d

25e

25f

26 广场外立面夜景

NIGHTSCAPE OF PLAZA
广场夜景

通过照明介入建筑，用照明再现建筑和从视觉信息方面附加建筑外立面更多内容，通过灯光艺术复现文化的历史和传承，以视觉性的再现帮助观者回想和发生联想。展现一座开放和具有无限活力的北国春城风貌，确定的灯光设计主题为"流光溢彩的电影长卷"，整个灯光的主题分为"黑白电影"、"彩色电影"、"数码电影"等章节来表达。强化建筑结构形式特点，营造出建筑夜景动态的秩序美，使得照明灯具与建筑结构高度一体化，做到真正的"见光不见灯"。

By means of lighting, the building is reproduced and endowed with more contents in terms of visual information on its facade. The light art presents cultural history and heritage, helps observers recall and associates, and demonstrates a spring city in northern China which is open and dynamic. The theme of lighting design is "a glittery film scroll" which is divided into chapters such as "black and white film", "color film", and "digital film". The lighting design consolidates structural form of the building, and creates dynamic beauty of order at nights, so that the light fixtures and building structural height coordinate in a harmonious way as if bright light is seen while fixtures are invisible.

27

28

27 一号入口夜景
28 二号入口夜景
29 浮雕夜景

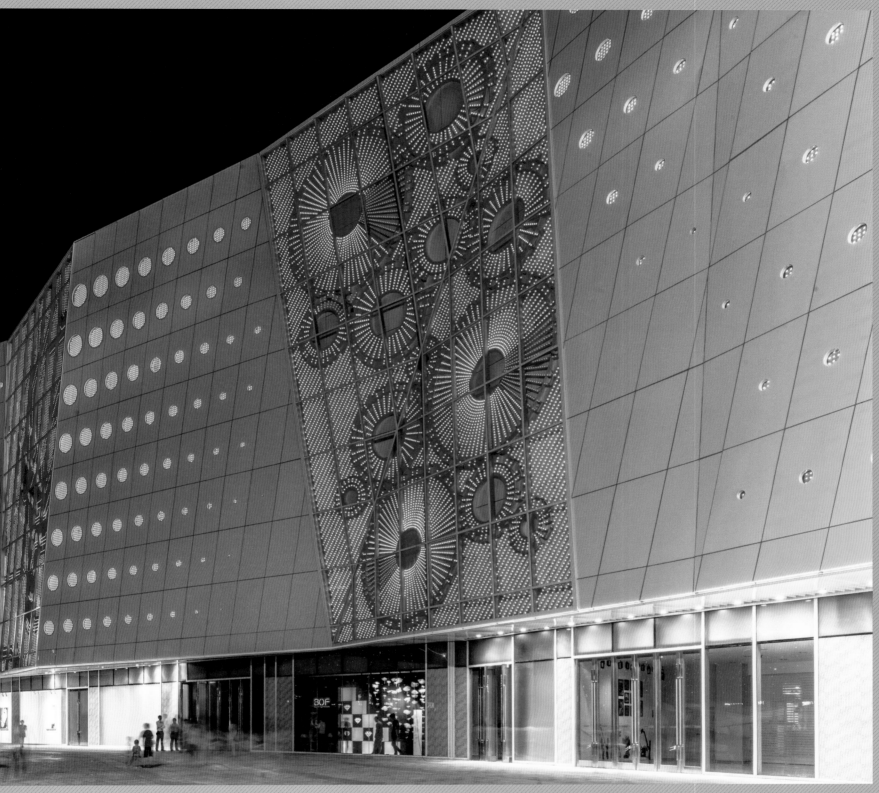

OUTDOOR PEDESTRIAN STREET
室外步行街

室外步行街的立面由多种不同形态的单元穿插组合而成，并由两条折线雨篷将整条街串起来，雨篷由商铺一层顶过渡至二层顶，形成一系列楔形的构筑物，构筑物为不同的材质和线条，丰富了整条街的形态。雨篷底部为密拼的百页，使室外步行街近人尺度的感觉非常细腻。

The outdoor commercial street of Kuancheng Wanda Plaza is composed of alternative units of various forms, and the awning of two folded strokes lines the whole street. The awning descends from top of GF to top of 2F, forming a group of wedge-shaped structures which are made of different materials and lines and enrich the content of the whole street. The bottom of the awning is tightly arranged blinds which create quite delicate sense of scale.

30

30　室外步行街门头
31　室外步行街日景
32　室外步行街入口
33　室外步行街夜景

HARBIN HAXI
WANDA PLAZA

哈尔滨哈西万达广场

开业时间 2013 / 09 / 13
建设地点 黑龙江 / 哈尔滨
占地面积 18.65 公顷
建筑面积 86.36 万平方米

OPENED ON SEPTEMBER 13 / 2013
LOCATION HARBIN / HEILONGJIANG PROVINCE
LAND AREA 18.65 HECTARES
FLOOR AREA 863,600 m²

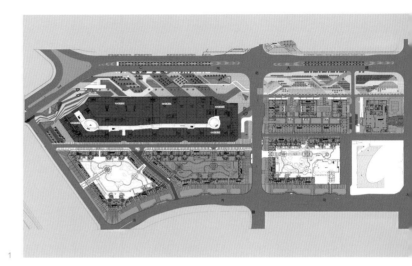

1

2

OVERVIEW OF PLAZA
广场概述

该项目位于哈尔滨市中兴大街南侧，南兴街以北，西宁南路以东和哈西大街以西。规划总建筑面积约86.36万平方米，其中商业13.41万平方米，五星级酒店4.09万平方米，5A高档写字楼3.85万平方米，写字楼9.75万平方米，商务酒店0.75万平方米。

建筑整体形象大气磅礴，设计手法融入哈尔滨当地文化特质和城市性格特征。商业部分着重打造整体的视觉冲击力，以三角形体块转折为基本要素，在强化商业氛围的同时实现群体建筑的协调统一。商业部分整体色调以白色为主，材质为异型铝板，形成明亮雅致的整体色调，体现商业氛围以及亲和力。

This project is located on the south of Zhongxing Street, north of Nanxing Street, east of Xining South Road, and west of Haxi Street in Harbin. The planned total floor area is 863,600 m², of which the commercial part is 134,100 m², the five-star hotel is 40,900 m², the 5A office building is 38,500 m², general office space is 97,500 m², and the business hotel is 7,500 m².

The overall image of the building complex appears grand and splendid, and the local culture and urban characteristics of Harbin are vividly demonstrated through its design. The commercial part, with focus on general visual impact, uses triangle blocks in transitions as its basic element, and attempts to strengthen commercial atmosphere and realize harmonious integration of the building complex. Basically in white tone, the commercial part applies abnormally shaped aluminum plates to form bright and elegant tone and imply strong business atmosphere and amiableness.

1 广场总平面图
2 广场鸟瞰图

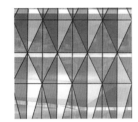
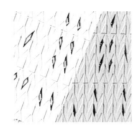

3

4

FACADE OF PLAZA
广场外装

营造沿中兴大道万达广场的商业氛围，形成统一大气的立面效果。设计手法以三角形体块转折为要素，着重打造建筑的视觉冲击力。大商业整体使用白色造型铝板，铝板细部衍生于冰花造型。以浅色调铝板、玻璃等为主要立面材料。商业主入口采用菱形玻璃幕墙，营造浓厚的商业氛围。

The exterior of Haxi Wanda Plaza is designed to create grand and commercial atmosphere along Zhongxing Street. The design uses triangle blocks in transitions as its basic element to build huge visual impact. Generally adopting white abnormally shaped plates from which ice flower patterns are arranged. The elevation materials include light tone aluminum plate and glass, and the main entrance applies rhombic glass curtain wall.

3 设计概念示意图
4 广场外立面
5 广场立面图

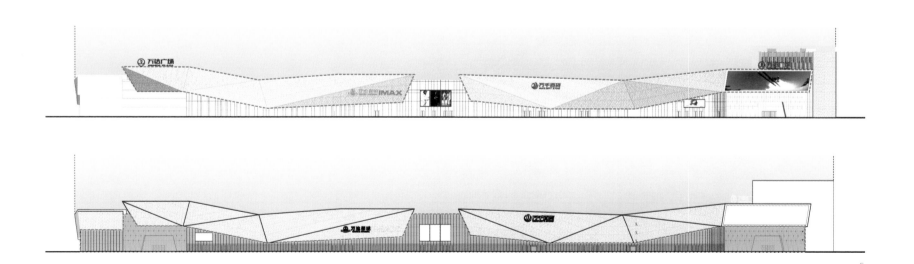

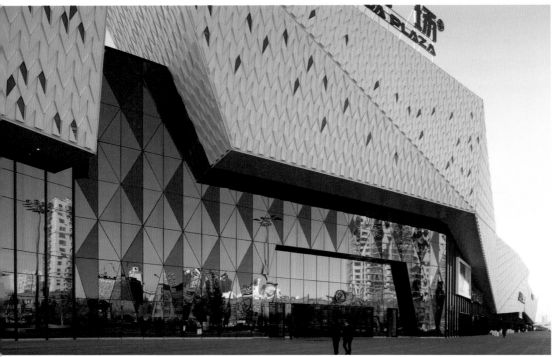

哈尔滨哈西万达广场的外立面设计源自哈尔滨地域
文化与万达企业文化的碰撞,设计将哈尔滨"冰城"
这一地域特点贯穿始终,无论是建筑形体的推敲还
是表皮肌理的处理,无处不体现"冰雪"这一灵动的
自然元素。

整个大商业以白色铝板为衬底,近1万块菱形白色铝
块均匀排布。表皮肌理运用三维立体白色铝板,以菱
形为基本元素,通过秩序排列且适当的点缀以散落
的点状玻璃,勾勒出"冰纹"的脉络。这种肌理同时
作为基本元素被延续到景观和室内空间,整个建筑
将冰雪飞舞的缤纷场景演绎得淋漓尽致,同时将现
实主义、抽象灵动、仿生自然的特色发挥到极限。

The facade design for Harbin Haxi Wanda Plaza is
originated from the interaction of Harbin's local culture
and the corporate culture of Wanda. The regional
feature of being an "ice city" of Harbin links the project
from start to end, and even the building shape and
treatment of building surface embody the natural
element of "ice and snow".

The whole commercial part evenly places around
10,000 pieces of rhombic white aluminum blocks as
its base plate. The surface applies 3D white aluminum
plates also basically in rhombic shape to make the
"ice veins" on the glass. This kind of veins, as the
basic design element, also extend to the landscape
and interior space, making the whole building a snowy
scene and playing the realism, abstract and bionic
nature to the extreme extent.

6 百货外立面
7 二号入口
8 外立面机理材质

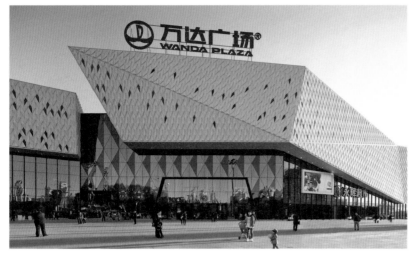

9

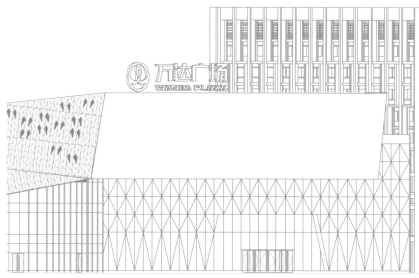

10

建筑的重点部位处理不仅细致, 而且富有张力。两侧入口部位以菱形玻璃为构成元素通过颜色变化呈现立体玻璃的效果。为了强化入口感, 在入口上方出挑61米×13.9米的巨幅广告灯箱, 与北面的哈尔滨西站遥相呼应, 抓点玻璃幕墙后衬巨幅滑雪图案, 气势恢宏, 效果极为震撼。

Major positions of the buildings are delicately dealt with full tension. The entrances on both sides are composed of rhombic glass to present the 3D effect through color changes. In order to strengthen the sense of entry, a huge advertisement light box of 61m×13.9m is overhead above the entrance, corresponding to the Harbin West Railway Station on the north. The dotted glass curtain wall is lined with a huge ski pattern to show its grandness and sensation.

11

9 二号入口
10 主入口立面图
11 一号入口

INTERIOR OF PLAZA
广场内装

整体空间设计以白色调为主，装饰造型理念由冰凌结晶构思而来，结合哈尔滨冰雕文化，配合暖色的灯光，将冰雕艺术文化体现得完美无暇。总之，其起伏的界面和空间，结合建筑形成富有动感的流线，营造具有活力的空间交替变化，是贯穿整个步行街广场的中心理念。无论是在步行街哪一处都能寻找到哈尔滨独特的地域文化，雪白的色调、穿插的动感，都与哈尔滨的独特文化气息完美地结合在了一起。

Generally in white tone, the interior space of the plaza is designed based on the ice crystal concept. In combination of ice sculpture culture of Harbin and warm light source, the ice sculpture art is perfectly presented. The undulant interface and space, together with the dynamic flow of the building, make a vigorous and alternative space, and becomes a central concept of the whole commercial street plaza. No matter where you are at the commercial street, you can always find the unique regional culture of Harbin, sense the white tone and dynamics in the perfect combination of the design and local culture.

12q

12 入口天花
13 圆中庭

12b

13

圆中庭主题元素的表现淋漓尽致，其观光电梯菱形隐框玻璃幕墙配合侧裙板圆润、温和的菱形底纹，与室外主体建筑形成内外呼应。

The theme element of the round atrium is in an incisive way. The rhombic concealed glass curtain walls of the panoramic lifts are equipped with side

椭圆中庭在设计过程中运用凸出的超白色背漆玻璃和连续的三角形图案组合，犹如开河的冰块；边上凹凸起伏三角形底纹，犹如江水般将冰块浮于表面。观光电梯的外立面，以蓝色贴膜玻璃构造而成，运用形式与中庭立面相呼应，但却如蓝色宝石般熠熠生辉，为整个中庭及整个广场平添了尊贵、优雅的气质，也使这座万达广场增添了一笔亮丽的色彩。

In the design of the oval atrium, it employs convex super white glass and continuous triangle patterns to imply ice cubes; the fluctuant triangle veins look like floods to support the ice cubes. The facade of the panoramic lifts is made of blue filmed glass which is glimmering like a blue diamond, endowing noble and elegant quality to the whole atrium, and also colorful brightness to Wanda Plaza.

15

14 椭圆中庭
15 椭圆中庭采光顶

16 连桥
17 室内步行街采光顶
18 室内步行街

长廊段的采光顶间隔地布置了圆点图案的彩釉玻璃，打破过长的步行街产生的单调感。侧裙板造型交叉、动感，运用三角形造型穿插其中。针对直街狭长的特点，为了避免给人造成视觉疲劳，通过桥体不同的形状、材质变化，使直街在统一中富于变化。

The skylight of the gallery is dotted with colored ceramic glaze which breaks the monotony of the super long commercial street. The side plates are intercrossing and dynamic with triangle blocks being placed. As for the narrow and long gallery, in order to avoid visual fatigue, the design applies different shapes and materials for the bridges to enrich the straight gallery space.

17

18

LANDSCAPE OF PLAZA
广场景观

整体概念贯穿哈尔滨冰雪世界的主题，与建筑、内装、夜景等融为一体。景观元素包括雕塑、灯具、花池水景等，全部以三角形体为立体构成的基本元素。在广场的东西两端的入口广场上，分别矗立着以东北林海雪原为原型的大型灯树组雕以及彩色水车喷泉雕塑。沿大商业立面的带状广场上，设计了8盏与大商业主体交相辉映的冰棱雕塑灯柱——这是本项目打破万达传统道旗灯形式的创新。景观以超凡脱俗的纯白色和简洁利落的线条雕刻哈西万达广场大气、简洁明快的气质，与建筑、夜景构成冰城特有的文化气息；同时暖黄色的发光灯柱，又恰如其分地为广场商业氛围带来了丝丝暖意。

The design concept goes through the theme of snowy world of Harbin in terms of building, interior and nightscape. The landscape elements include sculpture, light fixtures and pond waterscape, all in triangle shape to build 3D effect. On the east and west entry squares stand a group of large light tree sculpture with prototype of the snowy forest in northeastern China and a colored fountain sculpture. On the strip square along the large commercial elevation, 8 ice sculpture lampposts are designed to echo with the main body of the large commercial area, and this is an innovative way to break through the traditional flag light of Wanda. With purely white and concise strips the design cultivates the special glamour of Haxi Wanda Plaza which is grand, concise and graceful, constituting special cultural atmosphere together with the building and its nightscape. Meanwhile, the lampposts in warm yellow bring sense of warmness to the business environment of the plaza.

19

20

21

22

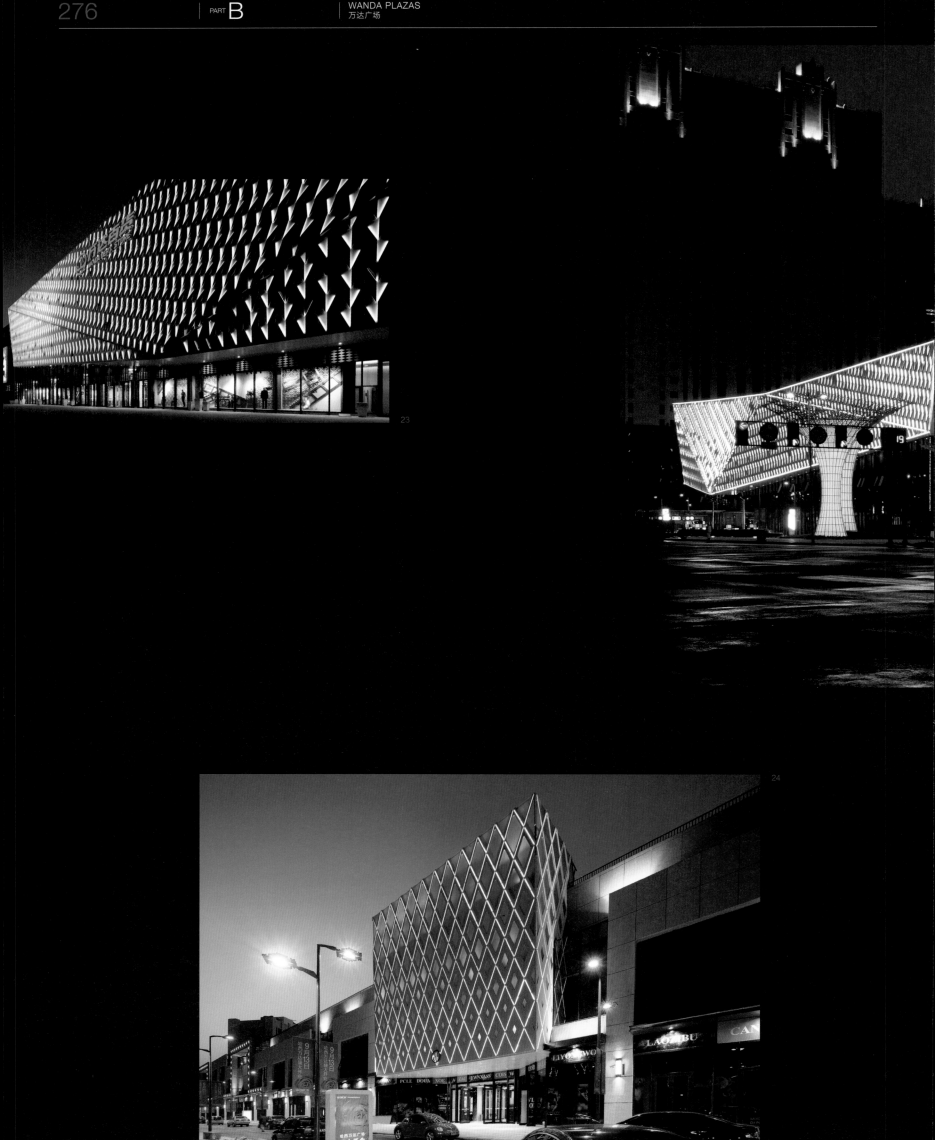

25

23 百货外立面夜景
24 门头夜景
25 广场外立面夜景

NIGHTSCAPE OF PLAZA
广场夜景

在整体灯光处理上，有效地融合建筑与艺术特色，全力塑造晶莹剔透、绚丽闪耀的冰雕艺术效果，突出体现"冰块"特色及"冰凌"的转折关系，特别是针对购物中心的造型特点，通过面的折叠关系，突出形体的变化，塑造钻石般璀璨的夜景效果。

In terms of general light treatment, the design effectively integrates with the architectural and artistic characteristics to present the glittering and glamorous ice sculpture art, highlighting the features and adversative relations of "ice cubes" and "icicles", especially for the shape characteristics of the shopping center, whose shape is strengthened through folding relations of faces to build glittering night effect.

FACADE OF HOTEL
酒店外装

哈尔滨哈西万达嘉华酒店位于大商业东侧地块。酒店总体轮廓呈简洁的竖向线条，高贵挺拔；外立面采用天然石材，典雅稳重；丰富细腻的线角，使建筑在转折、进退中增强了动感和光影的和谐变化。建筑气质充分展现万达积极向上的企业文化形象。

Wanda Realm Hotel Haxi lies on the eastern plot of the large commercial area of the plaza. The hotel profile applies concise vertical lines which appear tall and graceful. The facade in natural stone shows its delicacy and stability, and the rich line corners make the building more dynamic in the transition and advance, and realize harmonious changes of light shadow. The overall image of the hotel fully demonstrates active corporate image of Wanda Group.

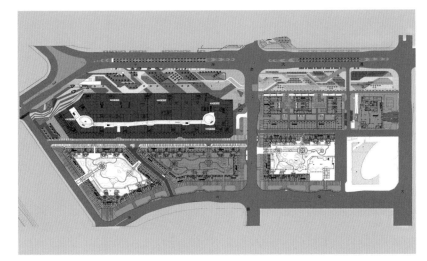

26

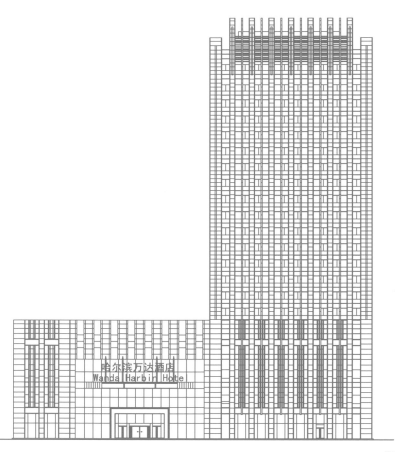

27

26 酒店总平面图
27 酒店立面图
28 酒店外立面

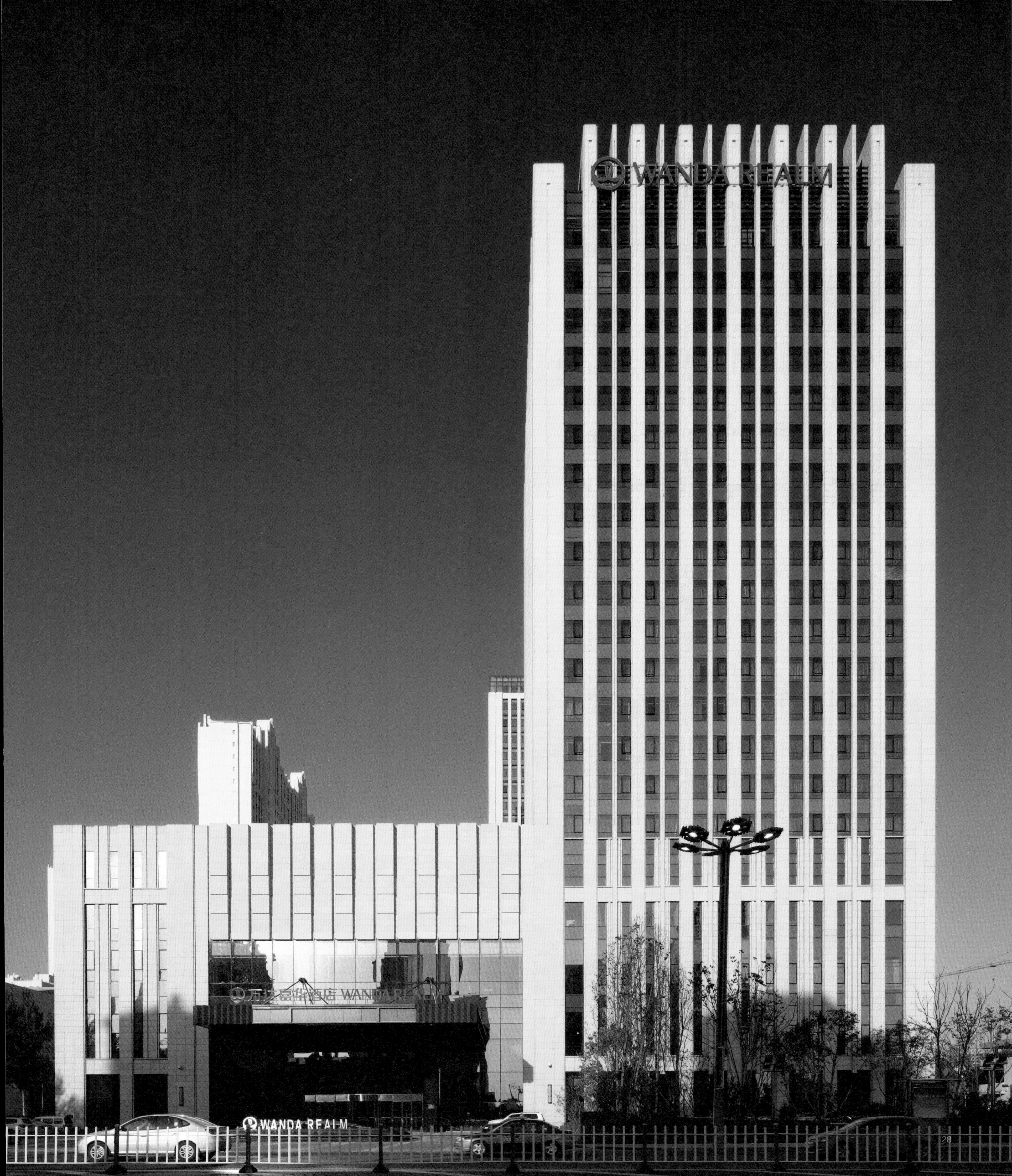
WANDA REALM
WANDA REALM

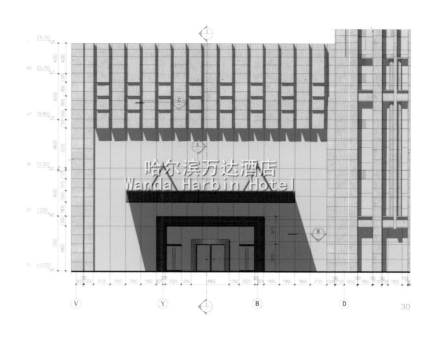

29

29　酒店入口
30　酒店入口立面图

银川万达嘉华酒店

北京

天津万达文华酒店

武汉万达嘉华酒店

南昌万达嘉华酒店

赤尾屿

钓鱼岛

台湾岛

东沙群岛

海南岛

台湾岛

海南岛

东沙群岛

西沙群岛

中沙群岛

黄岩岛

南沙群岛

曾母暗沙

南海诸岛

WANDA REALM WUHAN
武汉万达嘉华酒店

开业时间 2013 / 09 / 01
建设地点 湖北 / 武汉
客房数量 409 间
建筑面积 4.75 万平方米

OPENED ON SEPTEMBER 1 / 2013
LOCATION WUHAN / HUBEI PROVINCE
GUEST ROOMS 409
FLOOR AREA 47,500 m²

OVERVIEW OF HOTEL
酒店概述

武汉万达嘉华酒店总面积4.75万平方米，客房409
间。其中塔楼17层，每层面积1600平方米，总面积
2.72万平方米。酒店三层设有宴会厅、贵宾接待、会
议室4间和多功能会议室一间。其中宴会厅面积865
平方米，可容纳480人同时就餐。四层有泳池区、健
身房、美容美发和瑜伽室等。

Wanda Realm Wuhan has total floor area of 47,500 m²
and 409 guestrooms. The hotel's tower has 17 floors
with each floor area of 1,600 m² and total floor area
being 27,200 m², and its 3rd Floor is placed with a
banquet hall, VIP reception, 4 boardrooms and one
multifunctional conference room. The banquet hall
is 865 m² and can accommodate maximum 480
persons in total. A swimming pool, fitness center,
hair and beauty salon and yoga center are located
on 4th Floor.

1

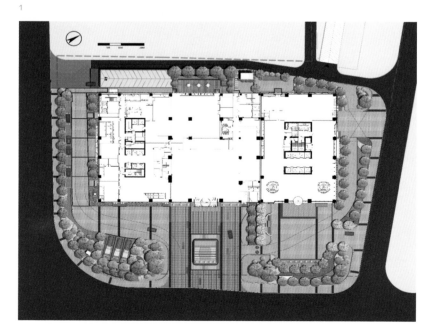

1 酒店总平面图
2 酒店外立面

2

FACADE OF HOTEL
酒店外装

酒店塔楼与裙房采用相同的设计手法,形成简洁、大方的立面效果。在透明玻璃和窗下墙玻璃幕墙基底,通过粗细相间的竖向铝合金装饰线条的转折及疏密变化,强调建筑底部的入口空间及体现建筑屋顶转角处的变化,创造一种富有动态平衡、具有灵活气氛的现代建筑风格。尤其酒店入口雨篷设计手法,更是巧妙地利用竖向线条的干净、利落地转折,很自然地将雨篷与建筑形成一个有机的整体。这些既整齐规律又富有变化的竖向铝合金装饰线条,与办公楼竖向垂直线条相呼应。酒店幕墙玻璃与铝合金材料的色彩均与办公楼一致,以到达整体建筑形象的协调统一。竖向垂直线条的间距及疏密变化根据酒店客房开间的尺度均匀分配,将客房使用功能的舒适度与立面分格尺度达到协调及统一。

The tower and podium of the hotel are designed with same method, which presents the simplicity and grandness of the facade. Vertical aluminum strips are alternatively placed on base of clear glass curtain wall and spandrel curtain wall, and the transition and changes of strips enriches the building entrance as well as the roof corners. In this way, a dynamic, balanced, modern and flexible building style is created. What is worth mentioning is the awning design at the hotel entrance, which skillfully utilizes clean and flexible transition of vertical strips to naturally form an organic integration of the awning and the building. These vertical aluminum strips which are in order but also dynamically correspond to the longitudinal strips of the office building. Colors of the hotel's glass curtain wall and aluminum alloy are the same as the office building, in order to indicate uniform coordination of the building image. Spacing and changes of the vertical strips are evenly allocated based on the division of each guestroom so that the comfort and scale of the guestrooms can reach balance and harmony.

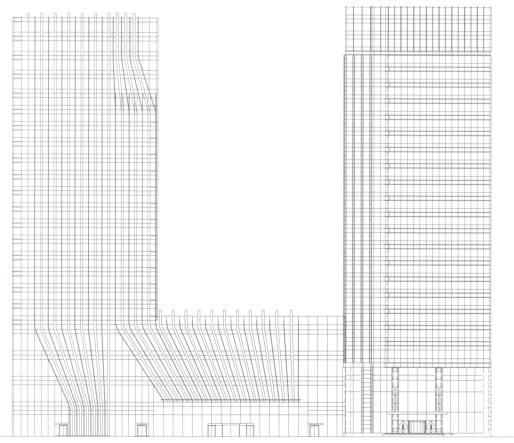

3 酒店外立面
4 酒店立面图

青山、瀑布、水流，婉转地隐现于建筑设计中。干净通透的青色玻璃幕墙作为整个设计的底色，将湖光山色依稀掩映。而附于其上的银色线条，疏密收分，从建筑顶部直落而下，最终收于酒店入口，并向前铺开形成雨篷——如同银河落泉，飞流直下，隐现出澎湃之势，使整个酒店入口尽现大气。而建筑四边各有线条的小收分，显得秀气端庄。各种建筑语言的应用使酒店在周围的建筑群中独具特色，也为湖岸两边增添了一道至美的风景。

Blue mountain, waterfall and stream are implied in the architectural design. The clear blue glass curtain wall,, is selected as the base tone of the whole design, vaguely showing off the scene of water and mountain. While the silver strips floating on the base are proportionally arranged from top to bottom and end at the hotel's entrance, forms an awning in the front of the hotel as if the Galaxy is falling at the entrance and making the entrance more grand and luxurious. The small reveals on four corners make the building look more grand and graceful. The application of different architectural languages makes the building distinct among the building clusters and endows the lake banks with gorgeous scenery.

5

LANDSCAPE OF HOTEL
酒店景观

武汉万达嘉华酒店位于万达武汉中央文化区，整个中央文化区占地120公顷。中央文化区以楚汉文化展开。景观设计师在设计酒店时紧紧围绕地块的整体文化氛围展开，并将楚汉文化延伸至酒店，从而使酒店与周围环境相融合。

广场前中心水景区域设计以突出汉文化大气、质朴的特质，展现汉代文化精华为要旨，水景平面布局由方和圆两种形式组合而成。中心跌水区与周围地面浮雕为方形；水景静水面区域中间放置圆球形特色镂刻景观小品，其镂刻图案与地面浮雕图案均设计成祥云形状，紧密结合设计主题。

Wanda Realm Wuhan is located at Wanda Wuhan Central Culture Area and occupies total land area of 120 hectares. The Central Culture Area is developed based on the Chu-Han culture, from which the whole culture atmosphere is designed for the hotel to coordinate with its surrounding environment.

The central waterscape area in front of the plaza is designed with grand and rustic features and express the essence of the Chu-Han culture. The layout of the waterscape is composed of square and round patterns, and the central drop area and its surrounding ground relief are square shaped. In the center of the static water face is placed with round featured landscape, and its carved patterns, together with the ground relief patterns, form auspicious clouds which are closely related to the design theme.

7 酒店水景特写
8 酒店喷泉特写
9 酒店喷泉
10 景观灯

10

NIGHTSCAPE OF HOTEL
酒店夜景

"疏影横斜水清浅，暗香浮动月黄昏"，虚实对比呈现，色彩时浓时淡，环境动静相宜，观景如梦如幻。使入住者顿感"弗趋荣利"、"趣向博远"的精神。

酒店立面纵向流畅的纹理，用正3000K色温的间接灯光贯通始末，酒店上部点缀或明或暗的星光灯，给刚刚体验了中央文化区一站式消费体验的旅客提供了宁静秀美、舒缓怡人的静谧空间。

"Water is clear and shallow to dredge the shadow crosswise, the dusk moon is floating over secret fragrance". The presentation of reality and virtuality, changes of color levels, coordinate to the static and dynamic of environment, as well as the dreaming land, all attract guests with higher spiritual pursue in the bustling reality.

The vertical flowing veins on the hotel facade are placed of positive 3000K color temperature indirect lights from start to end. On top of the hotel starry lamps are equipped, offering guests a tranquil, graceful and comfortable space with one-stop experiences in the central culture area.

11　鸟瞰夜景
12　酒店入口夜景

11　　12

14

WANDA VISTA TIANJIN
天津万达文华酒店

开业时间	2013 / 09 / 25
建设地点	天津
客房数量	297 间
建筑面积	4.85 万平方米

OPENED ON	SEPTEMBER 25 / 2013
LOCATION	TIANJIN
GUEST ROOMS	297
FLOOR AREA	48,500 m²

OVERVIEW OF HOTEL
酒店概述

天津万达文华酒店总面积4.85万平方米，地上层数22层，客房297间。首层为800平方米的挑空大堂、500平方米全日餐厅、415平方米大堂吧及150平方米红酒吧；二层为8间会议室，其中含1间200平方米以上的大会议室；三层设有会见厅和1400平方米的宴会厅；四层是酒店的康体设施，含游泳池、健身跳操和美容美发等功能。塔楼为客房层，含标准客房、套房、部长套房、总统套房等客房类型；19层为行政酒廊。塔楼顶部两层为会所。

Wanda Vista Tianjin occupies total floor area of 48,500 m², 22 floors above ground and 297 guestrooms. The hotel's GF consists of a doule-height lobby of 800 m², a 500 m² All Day Dining Restaurant, a 400 m² Lobby Bar, and a 150 m² Wine Bar; 2nd Floor has 8 boardrooms in which two are large conference rooms being over 200 m². 3rd Floor has a presence chamber and a 1,400 m² banquet hall. 4th Floor contains a swimming pool, fitness center, hair and beauty salon. The tower is for guestrooms which include Twin Rooms, Junior Suites, Ministerial Suites, and Presidential Suites. 19th Floor is an Executive Lounge, and the club house occupied the top two hotel floors.

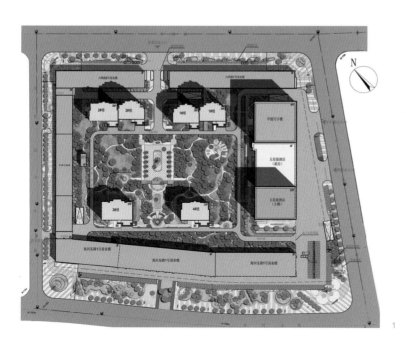

1 酒店总平面图
2 酒店鸟瞰

1

2

EXTERIOR OF HOTEL
酒店外装

天津万达文华酒店立面设计从城市历史和地理区位的角度出发,将Art-Deco这一传统建筑、装饰风格进行了转译与重构,运用当代新的材料、技术与工艺将传统与现代完美结合。整个立面设计贯穿"中西"、传统与现代的调和,演绎了经典的建筑横三段的基本格式。将古典与浪漫主义元素进行抽象与简化,并符合各体块之间和谐的比例尺度关系,高低错落叠置,形成极富现代感又颇具纪念性的造型。

The facade design for Wanda Vista Tianjin starts from the history and geographic location of Tianjin and reinterpret and restructure the traditional Art-Deco style by applying new modern materials, technologies and process in a perfect combination of tradition and modernity. It coordinates the "western and eastern" elements, tradition and modern factors to interpret the classic horizontal three-section form for buildings. It abstracts and simplifies the classic and romantic elements, conforms the harmonious proportion of different blocks at different height, and creates an extremely modern and monumental form.

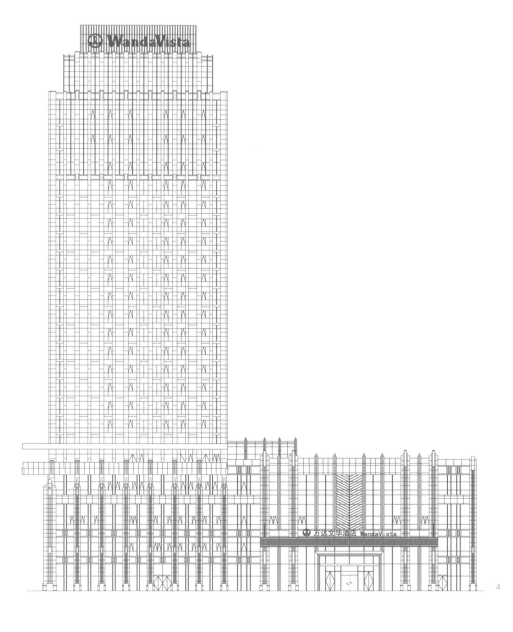

3 酒店外立面
4 酒店立面图

4

5

门头与入口幕墙处理比较简洁大气，较为通透，门套与雨篷关系结合紧密。雨篷天花采用钻石造型，配合灯光照明，彰显酒店高贵气质。

The gate frame and entrance curtain wall is concise and grand, creating a sense of transparency. The door pocket and canopy are designed in a close manner. Diamond-shaped canopy suspended ceiling and lighting jointly demonstrate elegance of the hotel.

5 酒店入口
6 酒店入口方案

7

7　酒店裙房外立面
8　酒店外立面细部

LANDSCAPE OF HOTEL
酒店景观

酒店景观具有开放性，视野通透，空间开阔。住宅景观具有私密性，道路曲折，绿植层叠。项目采用自然水系作为住宅小区与酒店的分隔，使酒店景观与住宅景观融为一体，互相借景，相映成趣，彼此衬托，自然和谐。

The hotel landscape is open and spacious. Luxury residential landscape highlights privacy, with zigzag roads and stacked green plants. While natural water system separates the luxury community and the hotel, it also blends the landscape of the two by borrowing scenery from each other, forming a delightful contrast and realizing natural harmony.

9

10

11

9 植物
10 雕像
11 牌楼

12 外立面
13 景观小品
14 灯柱

NIGHTSCAPE OF HOTEL
酒店夜景

夜景照明的高低错落模式呼应建筑造型，通过上、中、下不同层次灯光的营造为夜晚建筑的增辉，并能够充分表现出酒店典雅温馨。通过色彩的变化给人们带来强大视觉吸引的同时，在建筑立面边角位置形成竖向投光的"线"的照明方式与建筑各部位的面光形成呼应，留出充分的层次空间可以感受建筑的层次美。

Scattered lighting corresponds to the architectural shape. Lights at upper, medium and lower parts jointly create a well-crafted nightscape of the building and fully showcase elegance and warmness of the hotel. While bringing strong visual attraction to people, variety color helps to form a vertical casted "linear" lighting mode at corners of the building elevations to correspond to plane lights at all parts of the building, thus express sufficient space layering and beauty of the building.

16

15

15　酒店夜景效果图
16　酒店夜景

18

WANDA REALM NANCHANG

南昌万达嘉华酒店

开业时间	2013 / 12 / 13
建设地点	江西 / 南昌
客房数量	300 间
建筑面积	4.15 万平方米

OPENED ON	DECEMBER 13 / 2013
LOCATION	NANCHANG / JIANGXI PROVINCE
GUEST ROOMS	300
FLOOR AREA	41,500 m²

OVERVIEW OF HOTEL
酒店概述

南昌万达嘉华酒店位于南昌市红谷滩核心商务区，凤凰中大道与会展路交会处，是万达集团在江西投资的第一家豪华五星级酒店。酒店建筑面积4.15万平方米，地上20层，地下两层。酒店共300间客房，其中两联套客房13间、三联套部长套房3间、总统套房1间。酒店各类高档配套设施一应俱全，是地外赣江高楼商务区、高层生活区、高端办公社区的首家国际标准豪华五星级酒店。

Wanda Realm Hotel Nanchang is located at Honggutan central business district, the intersection of Phoenix Avenue and Huizhan Road in Nanchang City. It is the first luxury five-star hotel in Jiangxi Province that Wanda Group has invested. The hotel has floor area of 41,500 m², and has 20 floors above ground and 2 levels of basement. There are 300 guestrooms in total which contains 13 two-bedroom suites, 3 three-bedroom suites and 1 Presidential Suite. All kinds of exclusive facilities make the hotel become the first five-star luxury hotel of international standard of the high-end business district, high-rise living community and high-quality office area in Jiangxi Province.

1 酒店总平面图
2 酒店鸟瞰图

FACADE OF HOTEL
酒店外装

在酒店的整体设计构思中，着重强调了塔楼垂直通透的效果，通过材质的对比，创意结合功能，对大面积的玻璃幕墙和钻石造型的铝板构件作了精致的处理，使之优雅修长，同时也与一期的建筑造型呼应，建筑风格得以延续性，形成对话关系，从而将整个区域的万达建筑以一种和谐的方式统一起来。

塔楼的幕墙采用了钻石菱形的切割手法，以玻璃结合菱形铝板构件，彰显酒店大气却又不失精致的品牌形象。设计中通过对菱形铝板的交错布置，使立面更有节奏感和韵律感，同时也丰富了立面酒店转角部位的处理，使整个建筑更显新颖独特：通过对柱子的切角和折面玻璃的运用，使立面四边形成丰富的曲线变化，破除了普通塔式建筑竖直呆板的形象，呼应了主立面钻石的造型。

酒店塔楼顶部女儿墙，同样运用了与转角的处理手法，通过玻璃的折面，透影出仿如皇冠般的华丽，流光溢彩，气质非凡，彰显出酒店的高贵品质。

The overall design of the hotel highlights vertical transparency of the tower. Large-area glass curtain wall and diamond-shaped aluminum panels are finely designed by a comparison of material textures and a combination of innovation and function, makes them elegant and slim. In the meanwhile, they also correspond to the shape in the first phase to continue the architectural style and start a dialogue, thus harmoniously integrates all Wanda buildings in the region.

The tower curtain wall adopts the method of rhombic cutting of diamond-shaped, by integrating glasses and rhombic aluminum panels; it represents a grand while exquisite brand image. Staggered rhombic aluminum panels enhance the sense of rhythm of elevations, enriches the corners at the hotel elevations, thus making the whole building appear more novel and distinctive: the application of column chamfer and folded-surface glasses enable rich varieties of curves on four elevations, break up upright and ordinary tower treatment and echo with the diamond shape of the main elevation.

Parapet wall at the top of the tower adopts the same treatment method as the corners: folded glasses surfaces reflect crown-like magnificence, flowing light and color effect, and the manifesting noble quality of the hotel.

3　酒店外立面
4　酒店立面图
4

5

酒店入口挑出的雨篷，如大气的华盖，以简洁的横线凹凸收边，勾勒出精致外形。雨篷上部的双层玻璃则同样以菱形钻石为主体造型，白天漫射阳光，夜晚配合灯光闪动，剔透晶莹。

酒店裙房的设计通过多个菱形玻璃体的组合，形成一面钻石墙体，在环境光源的映衬下，美不胜收。通过对幕墙构造的分析，菱形幕墙为无框设计，保证还原出钻石折射的效果，成为本设计中的一大亮点。

Cantilevered canopy at the hotel entrance appears like a grand baldachin, with concave-and-convex edges in the form of concise horizontal lines, outlines a delicate shape. Double layer of glass on the upper part of the canopy also takes rhombic diamond as its main shape, diffusing sunshine at daytime and flashing light at night, shows an exquisite effect.

Multiple rhombic glasses on the hotel podium form a diamond-like wall, which is beautiful against the ambient light background. Based on analysis on the curtain wall structure, rhombic curtain wall adopts frame-free design to ensure diamond refraction effect, which is the highlight in the design.

5 酒店雨棚

6 裙房特写

7 酒店入口

6

LANDSCAPE OF HOTEL
酒店景观

酒店景观设计以赣江之水的元素作为整个景观设计的主题。水代表流动，代表改变，代表自我更新，代表智慧，代表坚持。在风水学说里常常用水来作为聚财的寓意。项目将流水的概念延伸到整个设计当中。铺装提取水波纹作为肌理元素，使流动的曲线贯穿场地，整个广场铺装具有强烈的流动感：水景的设计贯穿整个酒店前后场：前场主要做涌泉；流泉的主题水景设计，创造靓丽而极具识别性的特色景观。后场主要用一款静态优雅的镜面水景；静中有动，烘托宁静、优雅、宜人的气氛，营造人性化的休闲空间。

The landscape design of the hotel is themed by water elements of the Ganjiang River. Water represents mobility, change, self-renewal, wisdom and persistence. In Feng Shui theory, water often has the implied the meaning of wealth gathering. Flowing water concept is reflected in the whole project. By taking water ripple as texture element so as to make flowing curves run across the whole site, the Plaza pavement has a strong sense of mobility: waterscape design elements can be found at both the front and back areas of the hotel. To be specific, waterscape design in the front area is themed by flowing fountain, which creates a beautiful and distinctive landscape; while the back area mainly adopts a static and elegant reflective water feature. The combination of motion and quiescence helps to create a quiet and ambient atmosphere as well as a humanized leisure space.

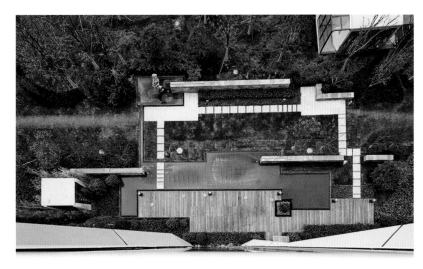

8 俯视景观
9 酒店景观绿化带

NIGHTSCAPE OF HOTEL
酒店夜景

夜景照明设计重在体现建筑本身的特质及纹理走向，提升建筑空间品质；力求将建筑夜景照明打造成区域的地标性灯光。酒店顶部运用400W钠投光灯打亮彩釉玻璃，釉面不少于60%，营造出金光璀璨的远观效果，提升建筑夜间的识别度。酒店塔楼立面LED洗墙灯旨在塑造、突出菱形铝板"钻石"立面造型；使酒店的整个楼体增强纵向感、挺拔感。

The nightscape lighting focuses on characteristics of the building and its texture orientation, with the aim to elevate the quality of architectural space and make the nightscape landmark of the region. 400W sodium project lights lit up colored glazing glasses (with no less than 60% ceramic) at the hotel top, create a shining effect when viewing from a distance, and makes the building more out-standing at night. LED flood lights on elevations of the tower create and highlight "diamond" elevation shape of rhombic aluminum plates, and makes the whole building apparently taller and uplifting.

11 外立面夜景
12 酒店入口夜景

WANDA REALM YINCHUAN

银川万达嘉华酒店

开业时间	2013 / 12 / 18
建设地点	宁夏 / 银川
客房数量	304 间
总建筑面积	3.89 万平方米

OPENED ON	DECEMBER 18 / 2013
LOCATION	YINCHUAN / NINGXIA HUI AUTONOMOUS REGION
GUEST ROOMS	304
FLOOR AREA	38,900 m²

OVERVIEW OF HOTEL
酒店概述

银川万达嘉华酒店项目位于银川市中心区，坐落于上海西路与亲水大街路口东南角，南侧比邻北京中路，东侧则是国际会展中心及人民广场西街。酒店部分总建筑面积3.89万平方米，其中地上3.24万平方米，地下0.65万平方米，304间客房；拥有与高档五星级酒店相匹配的配套设施——酒店大堂面积700平方米，挑空层高13米，首层还设置大堂吧、全日餐厅及日韩餐厅；二层为高档中餐厅；三层为宴会、会议层，包括一个面积为1200平方米的大宴会厅及多个大小会议室；四层为康体层，包括健身房、泳池和美容美发等功能用房。

The Wanda Realm Yinchuan is located in the central area of Yinchuan, at the southeast corner of West Shanghai Road and Qinshui Avenue, adjacent to Middle Beijing Road on the south and faces International Convention Center and West Renmin Plaza Street on the east. Covering a total floor area of 38,900 m² (including 32,400 m² for aboveground area and 6,500 m² for underground area), the hotel has 304 guest rooms. It is equipped with superior five-star hotel facilities: the hotel lobby is 700 m² in area and the lobby void space reaches 13m in height; the ground floor is set with lobby bar, three-meal restaurant and Japanese and Korean style restaurant; the second floor is arranged with high-grade Chinese restaurant; the third floor is for banquet and meeting, with a 1200 m² large banquet hall and several meeting rooms of different sizes; and the fourth floor is for fitness and recreation, including fitness room, swimming pool, cosmetology & hairdressing and other functional rooms.

1　酒店外立面

FACADE OF HOTEL
酒店外装

银川万达嘉华酒店项目所处的位置十分显赫——位于国际会展中心、行政中心、人民广场共同组成的城市景观轴线的尽端——总图规划阶段即把酒店和甲级写字楼布置为对称的双塔形式。采用相同的立面设计手法形成的双子座建筑体量，进一步强化了这一城市轴线的序列感。在酒店和甲级写字楼的裙房与顶部，均采用简洁的尖券交叉阵列的手法，增强了元素的韵律感和造型的对称性，强化了城市双子座的形象。在该项目的立面设计中，设计师将简洁时尚的现代设计理念，与银川作为伊斯兰文化继承和发扬集中地的城市地域传统文化特征有机地结合在一起，从而从多家设计单位的十余个备选方案中脱颖而出，最终得以实施。

Situated at one end of the urban axis jointly formed by the International Convention Center, administration center and Renmin Plaza, the project enjoys a very prominent location. The hotel and class "A" office building are designed as symmetric twin tower at the preliminary planning stage. The two buildings adopt the same elevation design method to form identical twin-buildings, further intensify the sequence of the urban axis. Pointed arches at the podium and at top of the hotel and class "A" office building are arranged in a simple crossed way, enhancing the sense of rhythm of elements and symmetry of the shape, and highlighting the image of urban twin building. The elevation design scheme which organically blends simple and fashionable modern design concepts with traditional cultural characteristics of Yinchuan, as the city inherits and carries from Islamic culture, was the selected winner from dozens of competition entries for final implementation.

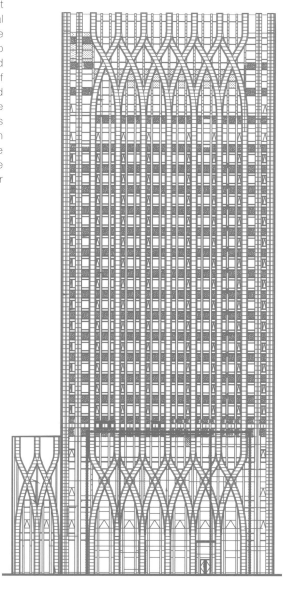

2 主入口雨棚
3 酒店立面图

3

在城市双子座之间的入口广场中心，设置了一个象征"银川之门"的建筑节点，作为酒店入口空间和城市公共轴线上的点景建筑。"银川之门"采用了拱券的形式，但却是将酒店建筑立面的拱券元素精细化使用。精细的石材民俗纹样雕刻，延续了当地的文化脉络，使得酒店前场室外环境充满了民族文化气息，并强化了双子座的形象特征。在藻井设计上，为了突出细部的材质和纹理，强化建筑的伊斯兰符号，做了图案的反复比选和推敲，最终使得"银川之门"能够完整地呈现出传统文化的美感和大气。

An architectural pavilion that symbolizes the "Gate of Yinchuan" is set at the center of the entrance plaza between the twin building, which serves as a landscape node for the hotel entrance along the urban public axis. The "Gate of Yinchuan" is designed as an arch, which is a kind of elaborate utilization of arch element of the hotel building elevation. Sophisticated stones carved with folk patterns are a kind of continuation of local cultural vein, adding ethnic cultural breath of outdoor environment in the front area of the hotel, and intensifying image characteristics of the twin building. Caisson pattern is determined by comparisons and discussions with the aim to highlight materials and textures in detail and to stress Islamic symbol of the building, thus makes the "Gate of Yinchuan" completely and showcase the sense of beauty and grandness of traditional culture.

4 雨棚特写
5 立面特写
6 藻井图案
7 银川之门

LANDSCAPE OF HOTEL
酒店景观

景观将伊斯兰建筑风格中的拱门与地面网格肌理和多边形花饰符号提炼出来，转化成景观空间和景观小品的营造手法，将景观空间与建筑交织融合，将伊斯兰花饰符号提炼融入景观构筑物、灯饰和小品中，形成银川万达嘉华酒店的特色，以鲜明的艺术性和浓厚的民族文化内涵展示自身的高端品质，并且与建筑风格融合，以达到酒店整体设计的统一。

By extracting arched door of an Islamic building style, and ground grid texture and polygonal ornamental design symbols, the meaning were all translated into landscape space and features. By extracting and incorporating the Islamic ornamental design symbols into landscape structures, decorative lighting and featured landscape, the project landscape gain its own characteristics, and demonstrates its unique high-end quality with its vivid artistic characteristics and rich of national culture connotation, and match the architectural style to realize a unified overall design of the hotel.

8

8　银川之门
9　建筑灯饰

11

10　绿化种植
11　导视灯柱

NIGHTSCAPE OF HOTEL
酒店夜景

酒店夜景照明，以展现高贵奢华为主，不但体现酒店自身的风格，还与周围的建筑群、环境等有所互动呼应，避免造成突兀与单调的感觉。在照明手法上，采用主光突出建筑的精彩部分，用辅助光带动其他部分，使楼体照明富有层次感，既突出建筑精华，又把建筑的整体形象表现出来。同时，严格掌握用光的方向，以表现建筑饰面材料的质感和特征，使建筑照明富有韵律和层次感。

Nightscape lighting design of the hotel is aimed to showcase its noble and luxurious features. It not only reflects its own style, but also corresponds to the surrounding building complexes and the environment, thus prevents it from giving people abrupt and monotonous feeling. Main lights highlight the design features of the building façade while auxiliary lights, and support other parts to make the layers of lighting effects, which showcases the overall building image while highlighting the building essence. In the meantime, light directions are well controlled to demonstrate textures and characteristics of building finishing materials, making the building lighting more rhythmic and hierarchical.

13

12 双子座外立面
13 酒店外立面

酒店入口夜景设计遵循建筑材质自身的特性，避免了在半镜面的拉丝不锈钢板结构上表现灯光，主灯具安装在菱形仿云石板内部形成发光灯箱效果，有效地利用透光云石板表现雨棚造型，营造奢华感。在凹槽处还采用线型灯来补充勾勒菱形结构，拉大了雨棚的进深感，进一步增强了视觉冲击力。

The hotel entrance nightscape design is made by considering characteristics of building materials, avoiding expressing light on semi-specular stainless steel drawing plate structure. The main lighting fixtures are built in rhombic scagliola plates to generate shining light box effect, and euphotic scagliola plates are effectively used in canopy shape so as to create a sense of luxury. Linear lights are set at grooves to outline the rhombic structure, widening depth of field of the canopy and further enhancing visual impact.

15a

15b

14 酒店入口
15 雨棚特写

DEVELOPMENT HISTORY OF INTERIOR QUALITY IMPROVEMENT OF WANDA INDOOR PEDESTRIAN STREET

万达广场室内步行街内装品质提升的沿革

文／万达商业规划研究院室内装饰所所长 毛晓虎

作为目前全球第二大商业不动产商,万达集团二十余年的成长历程印证了万达的确走出一条不同凡响、不可复制的发展之路,并借此使企业迈向了一个又一个的高峰。

万达广场室内步行街是一个贯穿整个广场的公共区域,是一条贯通整个广场的商业通道,更是万达商业广场的生命线。我们通过管中窥豹的方式,透过万达广场室内步行街内装设计的这一专项,来重温一下万达广场曾经走过的昨天,以及正在行进的今天和值得期望的明天。

2010年前,由于集团规模发展的需要,万达广场的建设速度决定了一切。18个月建设完成一个万达广场,16个月建设完成一个万达广场,万达集团在地产行业创造了一个又一个令人瞠目结舌的奇迹。当时,室内步行街的设计基于施工周期紧迫、施工条件不完备、核心设计理念不明确的情况下,所有设计的前提条件都是围绕着一个"快"字展开的,设计出图节奏快、材料加工快、施工进度快,室内设计风格也为配合这一系列的动作而趋于简洁,甚至于简单。这一时段的典型作品有:西安民乐园万达广场、上海周浦万达广场、青岛CBD万达广场等。我们不难看出,这些产品的室内设计有一个共同的特征就是:内装设计方案特征不突出,不强调室内商业氛围的创造,建筑设计与室内设计没有关联度(图1~图3)。

As the world's second largest commercial real estate developer, Wanda Group embarks an extraordinary yet non-replicable path of development over the past two decades, which underlies its growth to a new height.

Wanda Indoor Pedestrian Street is designed a public area which runs through the whole plaza, a commercial passageway connecting the whole plaza area and a lifeline of Wanda commercial plaza. Let us discuss what Wanda has done in the past, what it is doing currently, and what we can expect it in the future through the interior design of Wanda indoor pedestrian street.

Before 2010, due to demand of scale development, Wanda Group follows the speed-oriented principle in project construction. By completing construction of one Wanda plaza project within 18 or 16 months, the Group creates one after another astonishing miracles of the commercial real estate industry. At that time, most of its indoor pedestrian street design schemes were completed in the context of tight construction schedule, imperfect construction conditions and ambiguous core design concepts, and all designs were made in the premise of "high speed", with quick design drawing preparation, quick material processing process and quick construction schedule, and interior design style tended to be concise or even simpler so as to adapt to these circumstances. Typical projects completed by the Group during this period include Xi'an Minleyuan Wanda Plaza, Shanghai Zhoupu

（图1）上海周浦万达广场

（图2）西安民乐园万达广场

（图3）青岛CBD万达广场

（图4）厦门湖里万达广场

（图5）宁德万达广场

这个阶段属于万达广场室内步行街内装设计的初级和探索阶段，它的特点是：工程优先，速度优先，设计配合。

随着速度决定一切的时代逐渐过去，强调品牌塑造和品牌形象成为重点，在经过2010年前快速规模发展之后，当外部条件不断发生变化，企业品牌越来越强大之后，2011年万达集团提出了品质年的口号。由此，万达广场如同被打开了的宝盒，呈现出了一系列缤纷多彩的画卷。

2011年、2012年的部分开业项目，由于受到整体招商条件的影响，同时为了人为制造室内步行街的商业氛围，室内步行街在这个阶段的设计多采取了刻意强化装饰的多样性手段：在主入口空间、扶梯底部、室内连桥等交通节点部位、两个主要中庭的侧裙板装饰、直街侧裙等部位大量运用彩色LED灯光和各种图案肌理的装饰手法；这样"花哨"的设计语言，在开业初期的确是达到了旺场所需要的"商业氛围"，但随着商管经营手段的不断加强，我们在设计阶段着意落下的笔墨，却成了后期管理经营最为诟病的所在。

因为这些貌似在规则之下的各种设计手法，其实不仅有"喧宾夺主"之嫌，夺走了商业店铺在步行街中应有的"主导地位"，而且极大地增加了内装的设计成本、施工成本以及后期商管公司的运营成本。这个时期以厦门湖里万达广场、宁德万达广场、芜湖镜湖万达广场为典型代表，从照片上不难看出这一时期的万达广场可以明显感受到室内设计氛围的咄咄逼人；同时可以看出，较之2010年前朴素简洁的万达广场，这个阶段万达广场的室内设计明显有矫枉过

Wanda Plaza and Qingdao CBD Wanda Plaza. It can be easily seen that these projects share one common feature, that is, less prominent interior design scheme, non-emphasis on interior commercial atmosphere creation, and irrelevancy between architectural design and interior design (Fig. 1, 2 and 3). As the preliminary and exploratory stage of interior designs of Wanda indoor pedestrian streets, the stage is characterized by project priority, speed oriented and design for coordination.

After the age when speed determines everything has passed, brand building and brand image become a focus. After a rapid development before 2010, the brand becomes stronger along with constant changes of external conditions, the Group define the year of 2011 as the year of quality. Since then, just like an opened jewel box, Wanda Plaza start to project a series of colorful pictures.

In 2011 and 2012, due to the impact of overall merchant conditions and in order to create commercial atmosphere of indoor pedestrian streets, diverse means for projects are adopted for indoor pedestrian street put into operation, in order to deliberately highlight variety of decorations, including utilization in large area of color LED lights and various patterns at traffic nodes of main entrance, escalator underside and indoor connecting bridge, side skirt panels of two main atriums, and side skirts panels on straight gallery. These "showy" design languages indeed created "commercial atmosphere" required by flourishing such streets at the beginning of operation, but once the commercial management and operation methods improves, these "over-decorations" often brings problems in later-stage management and operation.

The reason they cause problem were in that these seemingly rule-conforming design methods actually

正之嫌。所以在当年度开业的万达广场，室内设计效果呈现出一个比一个"绚丽多彩"的时候，我们也在不断地反思和质疑这种设计风格和手法。这个阶段可以被概括为：纵情设计，矫枉过正，进退维谷（图4～图6）。

经过了2011年、2012年度的开业项目积淀与反思后，2013年万达广场室内步行街的装饰设计效果逐渐呈现出完全不同的创新思路：多样化、地域化的设计趋势已渐露端倪。

2013年的万达广场室内步行街设计可以用"绚烂之极，归于平淡，重塑设计"来形容，随着万达集团品牌影响力的逐渐扩大和每一座万达广场的建设周期的合理化，加上规划设计的日臻完善、招商条件的逐步好转，室内步行街也由设计商业氛围，逐步转变为由"一店一色"所主导的商业氛围，通过总结和梳理在2011年、2012年开业项目上取得的经验教训，室内步行街的设计于2013年回到了设计的本源上来，除了满足"永远品质年"的要求之外，如何进一步提升内装设计品质，成了设计关注的重点。

首先，强调建筑一体化设计，内装设计的手法汲取和提炼建筑外立面的设计要素，强化建筑外立面和内立面设计感的延续。其次，强调在保证室内空间延续建筑空间形态的同时，内装设计通常会对诸如连桥、共享空间等步行街公共区域的空间要素提出合理的调整和创新；同时，进一步强化"一店一色"的深化落实。

从某种意义上讲，万达广场的室内步行街设计走到今天，似乎又陷入设计的"瓶颈"了，那么，什么才是万达广场的明天呢？王健林董事长又提出了万达广场的设计应该是"追求文化，永无止境"，强调设计是文化的延伸，每一个万达广场的设计，无论是建筑设计还是室内设计，归根结底是一个发现地域文化特色和通过设计语言融汇这种特色的过程，所以每

not only "overshadow the main design" by taking away the "leading role" of commercial stores in pedestrian streets, but also greatly increase the cost for interior fitting-out, construction and operation of commercial management at later stages. Representative projects completed during this stage include Xiamen Huli Wanda Plaza, Ningde Wanda Plaza and Wuhu Jinghu Wanda Plaza. We can see from pictures that interior design of Wanda plaza project during this period is over-decorative, even somewhat overcorrect compared to the previous plain and simple Wanda plaza projects completed before 2010. Therefore, we keep reflecting on and doubting about the design styles and methods of plaza opened in 2011 and 2012, which are characterized by "bright and colorful" interior design effect. The stage can be summarized as over-design, hypercorrection and contradictory (Fig. 4, 5 and 6).

Based on re-study the above problems, interior design of Wanda indoor pedestrian streets that opened in 2013 gradually reflect quite different innovation idea, and a diverse and regional design trend has emerged.

Wanda indoor pedestrian street design in 2013 can be described as "splendid to its extreme, back to subtle and reshaping the design" stage. As Wanda Group's brand influence gradually expands and rationalized schedules of Wanda plaza construction, additional with the ever-improving planning design and favorable merchant conditions, indoor pedestrian street design gradually transforms into commercial atmosphere dominated by "one store, one style" from commercial atmosphere creation. Based on a summarization and comb of experiences in designing for projects opened in 2011 and 2012, indoor pedestrian street design in 2013 returned to its origin, which is focus on enhancing interior design quality while meets the requirement of "quality year forever".

Firstly, emphasis is laid on integrated building design. By coordinate with the exterior façade design, interior design intensifies continuity of indoor and outdoor façade design. Secondly, while ensuring the interior design will continue the spatial building forms, the interior design rationally adjust and innovate spatial elements in such public areas as connecting bridge and shared space; and further intensified the concept of "One Store, One Style".

In a sense, it seems that indoor pedestrian street design today again falls into a "bottleneck", and then what is the future of Wanda plazas? President Wan Jianlin puts forward that Wanda plaza design should follow the principle of "pursuing culture in an endless way", and stress that design is a extension of culture. Each Wanda plaza project design, no matter for architectural or interior, is in essential a process of discovering regional cultural characteristics and integrating such characteristics into design through design language. Therefore, each Wanda Plaza should follow the design principle of culture priority,

一个万达广场的设计都应该是：文化先行，设计落实，百花齐放。

在2014年即将开业的万达广场中，室内步行街的设计管理，就是着力从两个方向展开设计指引工作的，落实属于万达广场独特商业属性的文化创新、深入发掘地域文化的同时，力求在体现文化方面进行创新，并通过设计本身的美学变化来提升整个广场的文化含量（图7~图10）。

design in place and seeking excellence in all aspects.

Indoor pedestrian street design for Wanda Plaza to be opened in 2014 is guided by two principles, one is to follow through cultural innovation which is a unique commercial characteristic of plaza projects and deeply explore regional culture, and the other is to innovate in culture expression and elevate cultural features of the whole plaza by the aesthetic changes through design (Fig. 7, 8, 9 and 10).

（图7）厦门集美万达广场

（图9）东莞长安万达广场

（图8）大连高新万达广场

（图10）南京江宁万达广场

DESIGN AND CONTROL OF WANDA PLAZA GUIDE SIGNAGE SYSTEM

万达广场导向标识设计与管控

文／万达商业规划研究院室内装饰所副所长 万志斌

导向标识系统也称为"导视系统"，它有信号、标志、说明、指示、预示等多种含义；它的基本功能是指引方向，重要的辅助功能是强化区域形象，是引导人们在公共场所活动的综合性公共信息系统。

万达广场作为集零售、餐饮、娱乐、休闲、酒店及办公等诸多功能于一体的综合性公共空间，它的特点为空间步行化、街道室内化、业态多样化。针对这些规划特点，如何实现消费体验的便捷、各种业态空间的自由转换、功能需求的认知甚至消费导向的潜意识引导等需求，就需要导向标识系统对此进行专业的规划与设计，并实施有效的管控。

一、万达广场导向标识设计体系概述

万达广场导向标识设计体系，首先承载的是公众在万达广场所既定的公共空间行动时所需要了解的信息，包括场所地点信息、服务功能信息、商家业态信息、行为提示信息等，是对商业主体动线的阐述。这一系列的信息表达和传递是以导向标识为载体，帮助访客在不熟悉的环境中作出行动判断并顺利到达指定目的地。其次这套设计体系还在形式上丰富了空间形态的构成，从专业角度上丰富了消费者的购物体验与消费过程。

结合万达广场建筑规划的设计特点，万达广场导向标识设计体系主要分为：户外导向标识系统、室内导向标识系统以及地下停车场导向标识系统三大部分。

1 户外导向标识系统

包括：广场主LOGO塔、停车场及出租车停靠指示牌、户外综合及人行指示牌（含室外金街）等（图1、图2）。

Guide signage system is also known as "way finding system". It has multiple means including signs, logos, description, indication and prediction. Its basic function is to guide directions, and its important auxiliary function is to intensify the regional image. The guide signage system is a comprehensive public information system which offers people guidance in public areas.

Being a comprehensive commercial complex, Wanda Plaza integrates retail, food & beverage, recreation, leisure, hotel and office functions, Wand plaza is featured by pedestrian-oriented space, indoor streets and diversified programs. In response to these features, professional planning and design on the guide signage system should be effectively controlled in order to meet the demands of convenient consumption experience, free transformation of various commercial programs, recognition of functional demands, and sub-consciously leading the consumption.

I. OVERVIEW OF WANDA PLAZA GUIDE SIGNAGE DESIGN SYSTEM

First of all, Wanda Plaza guide signage design system contains information that public needs to know when they are in a specified public space, including information on location, service functions, type of business and activity notification,and interpretation of the main commercial streamlines of Wanda Plaza. All these series of information expression and transmission take guide signage a media to help visitors make judgment in an unfamiliar environment and get to destination smoothly. Secondly, this set of design system enriches spatial form compositions, and enhances shopping experience and consumption deed in a professional manner.

（图1）户外标识牌体

（图2）户外标识牌体

WANDA COMMERCIAL PLANNING 2013
万达商业规划 2013 —— 持有类物业 上册
347

2 室内导向标识系统

主要包括: 室内大堂综合信息牌、室内吊牌、扶梯吊牌及立牌、电梯厅立牌、卫生间吊牌、消防疏散指示牌等各功能类牌体。

3 地下停车场导向标识系统

墙面、柱面及地面的涂刷系统,车辆导向标识,客流导向标识,垂直交通位置标识等牌体(图3)。

Combined with architectural planning features of Wand plaza, the guide signage design system mainly covers outdoor guide signage system, indoor guide signage system, and underground parking lot guide signage system.

1. OUTDOOR GUIDE SIGNAGE SYSTEM

The outdoor guide signage system mainly include sign boards at main Logo stand of the plaza, parking

(图3) 室内标识牌体

二、万达广场导向标识设计原则概述

导向标识系统设计的原则应包含: 点位规划、造型设计、信息传递等几大方面。

1 设置的位置——标识点位规划原则

通常设置于流线方向变化节点,须结合现场环境设置,保证标识醒目、易识别; 须保证访客视觉通透性,不得阻碍访客行进路线。设计的主要依据之一是,经万达各系统评审确认的"万达广场车流动线"设计图纸。

2 载体的形式——造型设计原则

造型设计是为信息设计承载的载体,这个"载体"的设计应该与建筑、室内装饰设计风格、景观环境、地域文化等多重因素相协调一致,并满足牌体本身的尺度及材质的要求; 不同的牌体的尺度要满足各功

lot and taxi queue area as well as comprehensive outdoor and pedestrian sign boards (including outdoor gold street) (Fig. 1 and 2).

2. INDOOR GUIDE SIGNAGE SYSTEM

The indoor guide signage system mainly include comprehensive information boards in indoor lobby, ceiling mounted signage boards, ceiling mounted escalator signages and stand-along signage boards, stand-type boards in elevator halls, signage boards of wc, fire exit indication boards and other functional boards.

3. UNDERGROUND PARKING LOT GUIDE SIGNAGE SYSTEM

The underground parking lot guide signage system include wall, column and floor paint system, vehicular path signage, passenger flow guide signage, vertical traffic location sign boards, etc. (Fig. 3).

能性的要求；牌体的灯光设置要满足功能与装饰两方面不同的技术参数要求，并与所在空间的光环境协调一致。

3 信息的传达——版面信息设计原则

根据万达既定的信息导引原则来集合空间相关信息内容和识别距离，并根据规范及牌体造型设计原则设定文字字体、文字尺度、色彩关系、图标的使用、照明方式等要素，实现认知、引导的连续性和可读性。

三、万达广场导向标识设计管控

1 设计全过程的要点管控

根据导向标识系统设计的原则，万达形成了各专项的管控要点，在导向标识设计的各个阶段，均会根据该阶段的工作重点，由万达商业规划研究院牵头，组织项目公司、商管公司、设计单位等参与方，从设计、工艺、后期营运及需求等多角度对导向标识设计进行全方位的管控；在形成多方讨论会签的基础上进行后续设计工作的开展。

2 完善的复盘制度

每一个万达广场在开业60天后，都要由万达商业规划研究院牵头，组织项目公司、商管公司、实施单位、设计单位等参与方，从设计、工艺、质量、后期营运及需求等多角度对已实施的导向标识设计进行全方位的梳理；在形成多方讨论会签的基础上进行调改，以到达导向标识设计的真正目的，并有效满足后期的经营需求。

这种复盘不但是对设计实施的有效管控，同时也为不断总结、完善设计要求及相关的设计标准修订提供了科学化的设计依据（表1、表2）。

3 科研成果的标准化

针对万达广场导向标识的设计特点，结合历年开业项目导视复盘的共性问题，万达商业规划研究院组织集团商管拓展部、工程部、招商中心等多部门，在原有《万达广场地下停车场导向系统设计规范》进行了修编，并新增了户外及室内两部分，共同汇编成一套完整的商业综合体导向系统设计规范。

此规范的发布，是万达广场导向标识系统第一次拥有了真正意义上完整的设计标准，为规划设计系统、项目公司、商管公司等多系统在各自管控阶段提供了规范的设计及评审依据；同时也将通过设计标准中对布点原则、材质、工艺等的规定，从成本、质量等多个方面将导视系统提升到一个新的高度，为万达广场品质的全面提升提供了更科学的设计管理依据！

II. OVERVIEW OF WANDA PLAZA GUIDE SIGNAGE DESIGN PRINCIPLES

The guide signage system design should meet the principles of key location planning, shape design and information transformation.

1. LOCATION—GUIDE SIGNAGE POINT LOCATION PLANNING PRINCIPLES

The guide signage are generally located at nodes where circulation direction changes, and must be arranged by considering its environment to ensure that signs are clearly visible and easy to be identified; permeability of visitor's vision must be guaranteed, not blocking visitors' advancing route. One principle creation is defined by "Wanda Plaza Vehicular Flow Outline" design drawing reviewed and confirmed by all departments of Wanda Group.

2. FORM OF CARRIER: SHAPE DESIGN PRINCIPLE

Shape is carrier of the information, and such carrier design must be in harmony with architectural and indoor finishing styles, landscape, regional culture and other factors, and meet dimensional and material requirements of the board itself; dimensions of different boards should satisfy all functional requirements. Lighting for such boards must meet both functional and decorative technical requirements, and match with lighting environment where the boards locates.

3. INFORMATION TRANSFER: LAYOUT INFORMATION DESIGN PRINCIPLE

Information related to space and identification distance should be determined based on the existing information guide principle of the Group, and such elements as fonts of texts, size, color relationship and utilization of icons should be determined based on specifications and board shape design principles, thus making recognition and guidance continuous and readable.

III. DESIGN CONTROL OVERVIEW OF WANDA PLAZA GUIDE SIGNAGE

1. CONTROL ON KEY POINTS IN THE WHOLE DESIGN PROCESS

According to the guide signage system design principle, Wanda Group outlines key control points for different special items. During each design stage of the guide signage, Wanda Commercial Planning & Research Institute will, based on priority of the corresponding stage, organize project company, commercial management company and design company to carry out all-directional control on the guide signage design in terms of design, process, later stage operation and demand, and conduct subsequent design works after countersigned by all departments involved.

2. IMPROVED REPLAY SYSTEM

After each Wanda Plaza project opened for 60 days, Wanda Commercial Planning & Research Institute

（表1）万达标识系统验证报告调整表

（表2）万达广场标识实施阶段现场验证会签表

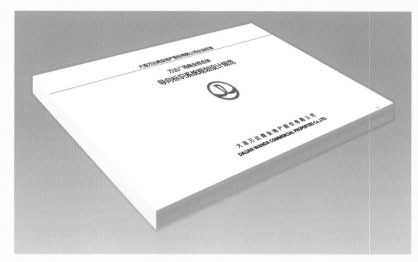

（图3）万达广场导向系统规划设计规范

正是基于这些设计标准的总结、研发，管控流程的建立与完善，万达广场的导向标识体系才在国家标准的基础上，结合万达广场的建筑规划、营运等特点，形成了属于自身特有的一套导向标识设计及管控的体系，不但有效地承载了人与建筑空间的互动，也成为万达广场高品质运营的有效保障！

will organize the project company, commercial management company, implementation company and design consultants to comb in an full-range manner that guide signage design already implemented in terms of design, process, quality, post-stage operation and demands, and adjust the design based on multiple departments countersign in order to realize the real goal of the guide signage design and effectively satisfy later stage operation demand.

Such re-evaluation system is not only a effective control measure for the design, but also offers a scientific basis for constant summary and improvement of design requirements and relevant design standards revisions (Table 1 and 2).

3. STANDARDIZATION OF SCIENTIFIC ACHIEVEMENTS

In response to the design features of Wanda plaza guide signage system, and based on common problems occurred during re-evaluation of projects opened in past years, Wanda Commercial Planning & Research Institute organizes commercial management expansion department, engineering department, investment promotion center and other departments to revise the *Code for Design of Wanda Plaza Underground Parking Lot Guidance System*, which supplement two chapters of the outdoor and indoor design, hence complete the whole set of code for commercial complex guidance system design.

The Code represents the first set of complete design standard on guide signage system of Wanda commercial complex, later named Wanda Plaza, and serves as the guidance in design and review of various departments (planning design system, project company and commercial management company) during their corresponding control stages; meanwhile, it stipulates key points allocation principles, materials and manufacturing, thus elevate the guide system to a new height in terms of cost and quality control, and offering more scientific design management benchmark for overall quality elevation of Wanda plaza projects!

Thanks to summary, R&D of the above design standards and establishment and improvement of control procedures, Wanda plaza guide signage system has formed its own design and control system based on national standard and by considering architectural planning and operation features of Wanda Plazas, which not only effectively enables interaction between people and architectural space, but also effectively guarantees high-quality operation of these projects!

DEVELOPMENT HISTORY OF WANDA PLAZA LANDSCAPE DESIGN

万达广场景观设计历史沿革

文／万达商业规划研究院总工办主任　王群华

"自2000年万达开始做商业地产以来已经走过了14个年头，从第一代店到第四代店，每一次升级都让万达跃上新的台阶"——节选自王健林董事长《商业地产投资建设》序言

万达广场每一个新的台阶都包含了景观品质的不断提升和超越，在不断发展的今天，不忘回顾万达商业综合体景观的历史沿革，为我们昨天的坚持留下厚重的影子，也为今天的创新提供可以超越的榜样。

万达广场规划模式经历了传统的单一盒子机构、多个盒子集合到盒子与建筑有机结合的过程，作为效果类的一个重要的组成部分，万达广场景观的发展也实现了从传统的大面积硬地铺装到绿化造景再到营造多元商业空间的发展历程。目前正朝着以文化主题的形式更具特色的方向发展，这也将使商业景观在辅助业态发展的同时突出自己的特性，不仅与商业业态相融合，而且要与其人文景观的魅力推动业态的发展。

一、硬铺时期

从2004年开业的南京新街口万达广场、武汉江汉万达广场及天津万达广场等代表的第一代产品开始，商业初期以沿街商业为主要表现形式。面积小，进深窄，景观以大面积硬铺为主要特点，以满足人流进出，缺乏满足人休憩停留的景观小环境。这一时期的景观严格意义上讲不能算真正的商业景观，这种状况一直延续到2009年左右（图1）。

（图1）南京新街口万达广场

"Since 2000, Wanda Group has been engaged in commercial real estate development for 14 years, evolving from the first generation Wanda Mall to the fourth generation store, each upgrading enables Wanda to scale a new high."—Quoted from the Preface of *Investment and Construction of Commercial Real Estate* written by Chairman Wang Jianlin

Each and every step Wanda Group has stepped on represents a process of constant improvement and transcendence of landscape quality. In today's continuously developing world, a review of landscape design development history of Wanda plaza not only reminds us of what we had striven hard for, but also offers an example to surpass today's innovation.

Wanda plaza planning mode witnesses various development stages, from traditional single-box structure, to a combination of multiple boxes and to an organic combination of blocks and building massings. As an integral part of visual effect design, the landscape of Wanda Plaza has developed from traditional large-area hard pavement to planting and versatile commercial space creation. Currently, Wanda Group is developing towards a more distinctive and culture-themed direction. In this sense, while supporting type of business development, commercial landscape will highlight its characteristics by not only integrating with commercial forms, but also promoting type of business development with its charm as humanistic landscape.

I. HARD PAVEMENT STAGE

The hard pavement period started from the first generation of Wanda plaza represented by Nanjing Xinjiekou Wanda Plaza, Wuhan Jianghan Wanda Plaza and Tianjin Wanda Plaza which were all opened in 2004. Retail along street is the main business model at early stage, whose stores were featured by small size, narrow depth and their landscape domained by large-area hard pavement. Such design only meets pedestrian circulation requirement but is lack of small greenary environment where visitors can stay for a relaxation. In strict sense, landscape at this stage is not qualified as commercial landscape, which situation lasted untill 2009 (Fig. 1).

二、大面积绿化时期

随着万达广场规模扩大和业态类型的丰富，建筑设计从单一盒子发展为多个盒子的集合体，商业景观设计随之更新换代，最突出表现特点是室外广场理念的出现，以北京石景山万达广场为代表（图2）。

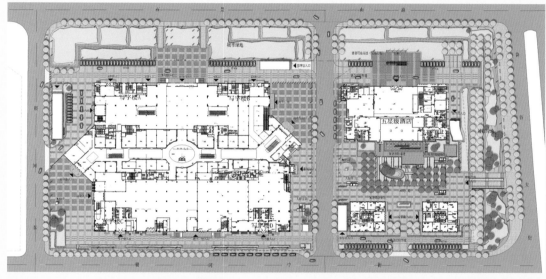

（图2）北京石景山万达广场

这一时期的景观出现了广场与非广场的功能区分，景观以绿化种植为主，讲究层次丰富、色彩多样的组合。从传统的硬地铺装为主，发展到以植物造景为核心的表现形式。植物种类增加，以大规格乔木配合灌木及地被的配置方式得到广泛应用，商业景观效果更具层次感。实现了人与自然、建筑与景观的有机结合，大幅提升了万达广场整体品质。但厚重的种植方式没有与住宅景观区分开来，并且不符合商业的功能要求。福州仓山万达广场和厦门湖里万达广场是这一时期景观的典型代表（图3、图4）。

II. LARGE-AREA GREENING STAGE

As sizes of Wanda Plaza, later named Wanda Plaza, grew and types of business diversified, architectural design has been developed from single-box to a combination of multiple blocks, and commercial landscape design also upgrade, in which the most prominent feature was the emergence of outdoor plaza concept. Beijing Shijingshan Wanda Plaza is an example (Fig. 2).

During this stage, plaza and non-plaza functions are distinguished. Landscape design is mainly realized through green plants, focusing on rich layers and a combination of different colors. This is a period with vegetation landscape as a core presentation form. Plant types are increased, allocation method of large-size arbor plus shrub and ground cover is widely applied,

（图3）福州仓山万达广场

（图4）厦门湖里万达广场

室外步行街的出现又把商业景观推向一个新阶段，小尺度的商业空间需要主题化更鲜明的情景式景观，通过提炼当地文化传统，进行艺术处理和加工，以时间或者故事发展脉络为情景式景观的联系纽带，将室外步行街的商业氛围烘托起来。

三、文化主题时期

在商业地产整个行业快速发展的背景下，商业景观体系的发展不断完善并调整思路，不断创新，更加注重特色文化主题和细节的表现。利用艺术化的雕塑小品为作为营造空间效果的主要元素，涌现出了泉州浦西万达广场、成都金牛万达广场、厦门集美万达广场、汉街万达广场、长沙开福万达广场等众多的优秀万达广场景观。直到这一阶段，万达广场景观规划体系的建立才基本完善，景观与城市功能、景观与建筑、景观与内装、景观与夜景照明等关系更加紧密，商业功能更加突出。景观设计与管控的思路得以充分释放，方法不断创新，精品项目特色鲜明，呈现出了"百家争鸣、百花齐放"的良性发展态势（图5~图8）。

四、多元购物体验氛围营造时期

随着儿童业态等娱乐业态的不断发展，人们的消费体验也从简单的购买商品逐渐发展到从购物体验中获取更多的消费需求。这样对购物环境的要求也随之提高，景观空间的塑造要能创造人与环境互动的

that makes the commercial landscape more layered, with an organic combination between human space and nature elements, and between building and landscape, which greatly elevates overall quality of Wanda Plaza. However, thick planting method is not distinguished from residential landscape area, and does not conform to commercial functional requirements. Fuzhou Cangshan Wanda Plaza and Xiamen Huli Wanda Plaza are two examples representative landscape design during this period (Fig. 3 and 4).

The emergence of outdoor pedestrian street push commercial landscape to a new level, in that small-size commercial space need more distinctive themed scenery, By refining local cultural tradition for artistic treatment and processing, and by taking timeline or story context as a bond linking to scenery landscape.

III. STAGE WITH CULTURE AS THEME OF LANDSCAPE DESIGN

With the rapid development of commercial real estate, commercial landscape system keeps on improvement and constant innovation, to adjust the focus on presentation of characteristic cultural themes, and the expression in details. Artistic sculptures are installed as main elements to create spatial effect, andmany excellent commercial landscape design projects emerged, including Quanzhou Puxi Wanda Plaza, Chengdu Jinniu Wanda Plaza, Xiamen Jimei Wanda Plaza, Wuhan Hanjie Wanda Plaza and Changsha Kaifu Wanda Plaza. Until this stage, Wanda commercial complex landscape planning system has basically settled, with more close relationship between

（图5）泉州浦西万达广场

（图6）成都金牛万达广场

（图7）厦门集美万达广场

（图8）长沙开福万达广场

场所；同时景观设施和元素要兼备文化性、娱乐性等要求，并能在不同地区、不同城市营造各有特色的购物环境体验。在万达广场景观规划体系中，不断丰富和完善景观空间和景观元素的参与性、文化性和娱乐性，不断丰富万达城市综合体的购物体验。

五、结语

万达广场景观体系的发展与万达商业功能的不断发展息息相关，商业功能中对综合体所处的城市、区域和内部的功能也会随着需求的变化而更加完善。商业景观的空间规划和元素设置也更加满足人在空间中的需求，人性化的景观是企业对社会责任的担当体现，更是景观品质化的最高境界。

landscape and urban functions, between landscape and architecture, landscape and interior, and landscape and nightscape lighting, which enhance the commercial functions. Landscape design and control are fully open-minded, innovative methods are adapted to present distinctive high quality projects, and showcase a sound development trend features "a hundred flowers blossom" (Fig. 5, 6, 7 and 8).

IV. STAGE OF CREATING VERSATILE SHOPPING EXPERIENCES AND ATMOSPHERE

As children and other recreation program developed, people's consumption experiences also shifted from simple commodity purchasing to seek of more consumptional demands through shopping experience, which puts shopping environment on a high requirement, and landscape space should be designed as a place facilitating an interaction between people and environment. Meanwhile, landscape facilities and elements must meet cultural and entertaining requirements, and be capable of creating unique shopping experience in different cities and regions. Comprehensive landscape planning system of Wanda Plaza should constantly enrich and improve the involvement, cultural relevant and entertainment, to enrich shopping experience for customers.

V. CONCLUSION

Wanda Plaza landscape system development is closely linked to continuous development of Wanda commercial functions, and programs of the complex are based on the city and region where the project located and the internal functions of the complex will be improved along with the demands changes. Spatial planning and element allocation of commercial landscape will satisfy people's special demands. Humanized landscape reflects the enterprise's social responsibility, which is the highest level of landscape quality.

2013
万达商业规划
持有类物业　　上册 VOL.1

WANDA COMMERCIAL PLANNING 2013
PROPERTIES FOR HOLDING

朱其玮 吴绿野 王群华 叶宇峰 兰峻文 张涛 黄勇
赖建燕 孙培宇 刘婷 张琳 苗凯峰 徐小莉 尚海燕
李文娟 安竞 马红 曹春 侯卫华 张振宇 范珑
谷建芳 张振宇 雷磊 王鑫 李彬 张飚 毛晓虎 莫鑫
都晖 刘江 蓝毅 郝宁克 屈娜 冯腾飞 张宝鹏 邵汀潇
万志斌 孙佳宁 袁志浩 阎红伟 吴迪 徐立军 王雪松
张立峰 陈维 谢冕 刘杰 党恩 高振江 沈余 孙海龙
李昕 李海龙 黄引达 孙辉 周澄 齐宗新 刘冰 潘立影
杨艳坤 程欢 邓金坷 康斌 刘易昆 李浩然 李江涛
钟光辉 张宁 张宇 黄春林 黄国辉 耿大治 刘阳
刘佩 石路也 孟祥宾 张洋 章宇峰 陈杰 冯志红
谷强 李小强 葛宁 张鹏翔 田中 虞朋 康宇 王治天
朱岩 董根泉 任腾飞 宫赫谣 王吉 沈文忠 张珈博
刘洋 胡存珊 马逸均 李光 郭晨光 朱迪 王锋 谢杰
李志华 宋锦华 方文奇 刘锋 秦鹏华 杨东 李涛
凌峰 易帆 华锡锋 任洪生 李明泽 刘刚 郭雪峰
陈嘉赟 孔新国 赵洪斌 刘志业 冯董 黄路 曹彦斌
张剑锋 周德 李易 肖敏 段堃 闫颀 唐杰 刘潇
黄川东 熊厚 张雪晖 董明海 李卓东 王静 王昉
谢云 李捷 关发扬 庞博 任意刚 张争 辛欣 傅博
赵陨 杨春龙 顾梦炜 姜云娇 江智亮 白宝伟 王凤华
李健 卫立新 庞庆 何志勇 宋永成 谭訸 卜少乐
高杉楠 韩冰 刘海洋 高峰 王睿麟 王宝柱 野天星
王瑶 葛朗 张佳 王晓昉 曹国峰 李常春 徐春辉
王永磊 于崇 张勰 杨汉国 王文广 张永战 李晓山
罗冲 张旭 高达 赵晓萌 方伟 刘俊 陈海亮 康冠军
晁志鹏 邹洪 郑鑫 周永会 陈志强 陈涛 张宇楠
张绍哲 刘安 全永强 康兴梁 林彦 路清淇 陈晓州
白宇 汤英杰 钱昆 白夜 崔勇 陈理力 刘昕 韦云
杨华 金柱 马辉 杨娜 王朝忠 罗琼 洪斌 刘晓波
赵宁 韩博 徐广揆 张烁君 金博 魏大强 程波 马骁
王鹏 柏久绪 朱广宇 蒲峰 杜晶晶 汤钧 冯科力
主佳 张浩 李扬 刘佳 青云富 王燕 田杰 熊伟 董丽梅
曾明 戎帅 陆峰 李峥 莫力生 李楠 屠波 王巍 吴昊
杨洪海 王惟 刘子瑜 李兵 李树靖 杨宜良 宋樱樱
刘冰 石紫光 牛晋华 程鹏 李为状 王力平 秦彬 王宁

（以入职先后为序）